Old Physics for New:
a worldview alternative to
Einstein's relativity theory

Thomas E. Phipps, Jr.

Apeiron
Montreal

Published by C. Roy Keys Inc.
4405, rue St-Dominique
Montreal, Quebec H2W 2B2 Canada
http://redshift.vif.com

First Published 2006

Library and Archives Canada Cataloguing in Publication

Phipps, Thomas E., 1925-
   Old physics for new : a worldview alternative to Einstein's relativity
theory / Thomas E. Phipps.

Includes bibliographical references and index.
ISBN 0-9732911-4-1

   1. Special relativity (Physics). 2. Electromagnetic theory. I. Title.

QC173.585.P52 2006   530.11   C2006-905979-9

Appendix quoted from: *A Threefold Cord: Philosophy, Science, Religion*, Viscount Samuel and Herbert Dingle, © 1942 George Allen and Unwin. Reproduced by permission of Taylor and Francis Books UK.

Front cover: A blue Fu dog (at Allerton Park, near Monticello, Illinois), symbolizing Nature or 𝕺𝕷𝕯 𝕻𝖍𝖞𝖘𝖎𝖈𝖘, regards with dismay the glittering **New Physics** of the "black Garuda" (a mask crafted by Balinese artist Nyoman Setiawan), one of various incarnations of Vishnu, in the present context seen as the incarnation of Albert Einstein's special relativity theory.

Table of Contents

Foreword

These few words of introduction are primarily directed at those readers who are not familiar either with Tom Phipps' singular style of scientific prose or, more importantly, with his rigorously applied view that, when theorizing about the world around us, we must pay absolute attention to the *practicalities* of the measurement processes by which the quantities involved in this theorizing are measured.

Let me talk about the Phippsian prose style first: the common experience upon reading a scientific text is to be confronted by a *finished article*—that is, by a text from which all sense of *intellectual journeying* has been exorcised, cleansed, deleted. The experience may be necessary but it is rarely exciting and never invigorating—it becomes merely a job that must be done, a dusty dry road along which weary feet must be dragged. But Phipps refutes this puritanical model; he is *renaissance* man—the man who glories in the splendour of the written word and its capacity to illuminate the obscure, and to decorate the plain. And so the experience of reading Phippsian scientific prose is not unlike that of reading a good detective novel—the dim detective, the obvious clues overlooked, the false trail followed, the unsolved crime written up as *solved* so that the bureaucrat can sleep his dreamless sleep and, finally, Sherlock Holmes with his pipe and Dr. Watson ...

Now let me consider the (for me) perfectly commonsensical view that the practicalities of the measurement process must play an unambiguously prominent role in the theorizing process: As an example of a theory where this was not done (with hugely significant consequences), we need look no further than classical Maxwell electrodynamics. In this case, the formalism absolutely requires that the detectors used by (inertial) observers to measure field quantities be *at rest* in the observer's frame. Thus, if we have two observers, each in his own inertial frame, then, since their instruments are physical objects and unable to occupy the same place at the same time, it is absolutely impossible for these two observers to make simultaneous measurements of the same field point. In other words, certain choices made at the theorizing level have rendered impossible a perfectly reasonable thing—that distinct observers can have direct knowledge of conditions occurring at a particular place at a given time. Phipps' answer to this conundrum is simple: there is no reason on Earth why the detector measuring field quantities should be fixed in the (inertial) observer's frame. After all, the source currents which generate the field are not, so why should the test-particles (which comprise the detectors) be? And since the detector need not be fixed in one observer's inertial frame, why should it be fixed in *any* inertial frame? Following this logic, if we allow the detector to have free motion, then the formalism of electrodynamics which follows must somehow allow for the *parameterization* of the detector's motion. A natural candidate for this formalism already exists in the equations of Hertz's electromagnetic theory (the known failure of his theory was the fault not of his equations but of his physical interpretation) and these are easily written down: just take Maxwell's equations and replace all appearances of $\partial/\partial t$ by d/dt. This replacement introduces a convective velocity which must be interpreted, and Phipps' solution is to use this convective velocity to describe the motion of the free detector. A simple and elegant idea, don't you think? ... but now comes the crux: by this simple process, which is driven by the idea that there is no reason on God's Earth why an observer cannot use a freely moving detector, the equations of electromagnetism become *Galilean invariant;* thus, at a stroke, solving one of the great conundrums of 19th century physics and, in removing the primary *raison d'être* of Special Relativity (SRT), putting a huge question mark over a large chunk of 20th century theoretical physics.

Now Phipps is a realistic and honest man and each of these traits has its consequence on the way his thinking proceeds. Realism first: the story outlined above makes plain that SRT, and all that has flowed from it, is an unfortunate accident of history for some and an incredible stroke of good fortune for others—and it is the 'others' who are in the driving seat here. What is required, Phipps realizes, is an example of some physical circumstance in which SRT can be shown to have failed ... *unambiguously.* One does exist, although careful reading of the standard texts (when one is wide awake and on top of one's game) is required to spot it, otherwise the cardsharp cleans you out: *stellar aberration* is the bone in the fish pie. Briefly, and as Phipps points out in entertaining detail, SRT *claims* to provide the complete explanation for the Doppler shift and for stellar aberration—both phenomena affecting light that comes from stars. To see the problem immediately, it is sufficient to observe that in order to explain the Doppler shift, the velocity used by SRT is defined as the relative velocity between emitter and detector ($v = v_e - v_d$) which, of course, is perfectly consistent with SRT's own internal logic. However, in order to explain stellar aberration, the velocity used by SRT is defined as the Earth's orbital velocity in the solar frame ($v = v_{orb}$) ... stellar velocities are nowhere to be seen ... and there is no source-sink *relativity* whatsoever! So, in order to 'explain' two different aspects of the same starlight, SRT must submit to two different interpretations, one consistent with its own internal logic and one *inconsistent* with that logic. If you work in a University physics department, try putting that position to any of your colleagues.

Honesty second: there are several good reasons for being extremely sceptical about SRT—Phipps is eloquent on them all—but he knows that the clock cannot be turned back to 1894 (the year Hertz died). Physics has moved on since then (and I do not mean merely *theoretical* physics); in particular, although we can with reason reject SRT, the *time dilation* prediction of SRT has been verified to high accuracy many times over. Indeed, without using the time-dilation *formula* of SRT to calibrate the relative clock-rates of the Earth-based clocks and orbiting clocks, the GPS system could never work as well as it does. So, Phipps accepts that *time dilation* is a fact of physics and that the time-dilation *formula* of SRT is verified and must therefore be properly built into theory.

So, how does Phipps respond to this state of affairs? Well, close analysis is hardly required ... for the Emperor is clearly naked to the innocent eye ... SRT makes *two* independent statements, of which we are all aware: firstly, there is the statement about *time dilation* (with a formula which works in well-defined situations) and, secondly, there is a statement about *length contraction* ... which Phipps correctly points out is a prediction of an effect which (a) has never been observed and (b) creates all kinds of difficulties, not least of which is making it impossible to consider SRT as a generalization (or *covering theory* in Phippsian lingo) of Newtonian Mechanics. It is the *length contraction* prediction, for example, that makes the science of rigid body mechanics *impossible* for the "relativist." For Phipps, and for any right-thinking person in my view, the notion of length-contraction is a metaphysical fantasy that can have no place in a theory of *physics*. And because *length contraction* and *time dilation* are independent statements then—as Phipps points out—we can cherry-pick. We *can* have a theory which assumes the reality of time dilation whilst denying that of length contraction. The way forward is formally trivial—just replace ordinary clock time, t, in the Hertz formalism by the proper time parameter, τ, defined in the usual SRT way where the velocity parameter, v_d, is the velocity of the detector in the (inertial) observer's frame. The result is the *Neo-Hertzian* formalism, the ramifications of which Phipps works through in great detail—but I shall stick with the big canvas: in denying the existence of *length contraction* but accepting the existence of *time dilation* Phipps is, in fact, denying *spacetime symmetry*; but, in doing so, is regaining the possibility of rigid-body mechanics *and*, through the neo-Hertzian formalism, is finding mutually consistent treatments for the Doppler shift *and* stellar aberration. This is already a huge bonus.

This Phippsian saga closes with a couple of chapters devoted to discourse on the nature of timekeeping (rather than on the nature of time). As I see it, this section is driven by three circumstances: firstly, there is no identifiable *causal* mechanism within SRT for the "predicted" *physical effect* of clock retardation. If there were, the twin-paradox would never have arisen in the first place. Secondly, there is the (almost) self-evident fact that any man (or, in this politically correct world, person) engaged in theorizing about the world armed with a sensibly constructed clock which furnishes a time t, can either choose to use t directly as his measure of time *or* choose to use an arbitrarily defined strictly monotonically increasing function $T = g(t)$ as his measure of time. The only consequence is that there will be some choice of g which provides maximal simplicity to his theorizing—*but all choices are equally valid*. Thirdly, there is the empirical fact of the engineers' experience about how to make the GPS system work in practice—the fact of an Earth-bound Master Clock against which all the to-be-launched satellite clocks are calibrated so that *once they are in orbit they keep synchronous time with the Earth-bound Master Clock*. This calibration process amounts to the choice of a set of g-functions $-g_1, g_2,...$ say—each one tailored separately to account for the distinct operating conditions of its associated clock. In effect, Phipps argues that there are no reasons whatsoever—

beyond vain prejudice and ideology—for believing that there exists *for any system* an inherently fundamental measure of time (or "proper time" in the sense intended within SRT and GR). And, upon reflection, I find myself agreeing with him. In which case, he argues, the most simple system of time-keeping is the one pioneered by the GPS engineers—that of an agreed (almost inertial) Master Clock against which all other clocks placed wheresoever are synchronized by a *g*-transformation chosen according to the operating conditions of the clock concerned (gravitational potential, relative velocity, *etc.*, accounted for). Thus, the vision spawned by SRT & GR according to which there are as many different "proper" clocks as there are particles in the universe is replaced by one in which there is a single (arbitrarily chosen) inertial Master Clock against which all other clocks are synchronized. As always, Phipps provides an exhaustive analysis of the ramifications of this timekeeping methodology—but two can be mentioned in a single breath: the absoluteness of the *here and now* is restored to the discourse of physics—with the corresponding consignment of the *relativity of simultaneity* to the proverbial dustbin; and the resurrection of the distinct possibility of a realistic theory of many-particle physics.

I shall finish my few words in praise of this lovely, lovely book by remarking briefly on that aspect of the neo-Hertzian formalism which I find to be most remarkable: as a student (forty years ago) I struggled with Maxwellian electrodynamics, and part of my problem was that I always found two things rather odd: firstly was the fact that here we had a theory in which the (supposedly) most important parts were the fields, **E** and **B**, which were unashamedly defined in terms of Newtonian forces—and yet this very same theory was proclaimed the fountain-head of all that was non-Newtonian in the whole world; secondly was the fact that, although ideas of *force* were hard-wired into the definitions of the field quantities, the theory still required an additional *postulate* (the Lorentz force law) to make it into a useful theory of electrodynamics. One can accept such things in an entirely mechanical way, of course. But they left me feeling perpetually slightly disconnected from any claim to a real understanding of the Maxwellian picture. At a stroke, Phipps has removed all such impedimenta to clear sight: no longer is electrodynamics claimed as the portal to a shining new world, quite different from the old; instead, it sits firmly and squarely as an integral part of that old world. And, almost by magic—yet not really—Phipps shows us that, in its neo-Hertzian reincarnation, electromagnetism is already *electrodynamics;* there is no need to postulate force laws additional to those inherent in the basic definitions of the field quantities ... read, marvel and enjoy!

David Roscoe
Sheffield, October 2006

The great discoveries of science often consist ... in the uncovering of a truth buried under the rubble of traditional prejudice, in getting out of *cul-de-sacs* into which formal reasoning divorced from reality leads; in liberating the mind trapped between the iron teeth of dogma.

—Arthur Koestler, *The Sleepwalkers.*

Author's Preface

Science can be seen as the fitting to nature of a mask—a mask beneath which the probings of human curiosity may be able to discover only other masks. For nature, as the ancients knew, is *something else*. Some spectators today, sufficiently concerned to qualify as "dissidents"—of whom I happen to be one—view the mask currently popular among professional physicists as especially and unnecessarily hideous. Beauty is in the eye of the beholder ... and ugliness, too. A keen sense of the ugly is as essential to the success of a would-be contributor to science as any of the more obvious talents. For the points of special ugliness of a theory are the *loci* of its vulnerability, at which a sharply-directed attack is most likely to lead to progress.

In the human genetic mix the lovers of ugliness seem to be a majority dominant in every era. By them, ugliness is perceived as beauty ... and there is little arguing with them, since they command, as needed, all the machinery of political dominance, including both democracy (the voice of consensus) and autocracy (the voice of authority). Consequently, identifiable "progress" is a matter of accident, a statistical fluctuation against odds. Old ideas pass out of fashion, old beauties succumb to new. In the seething ferment of transiently fashionable ideas that is the frontier of today's theoretical physics, the only trend visible to the eye of the detached, not to say bemused, beholder is a secular increase of ugliness—as of entropy.

Heading the parade of modern physics ugliness is Special Relativity Theory (SRT), an icon now so sacred that to breathe a word of negative criticism is to be automatically awarded the jester's bells and mantle of "crackpot." Many critics (misled by the tale of "The Emperor's New Clothes") have tried to laugh, expecting to evoke a chorus ... and all such have left their bones to whiten the Juggernaut's path. I anticipate no different fate, and am not concerned except to leave one more set of footprints on the path.

A hallmark of ugliness is pretentiousness, and few ideas in the history of physics have been more overweeningly pretentious than the one underlying SRT, that of Universal Covariance. Maxwell's equations are ugly enough, being invariant not even at first order ... but Einstein and Minkowski had the inspiration to universalize this form of ugliness and make it beautiful by making it transcendent — over mechanics and all the rest. The present investigation is concerned with showing ways in which that particular mask fails to fit nature. Though a goodly portion of this book will be devoted to destructive criticism — for ugliness deserves no less — most of it will be concerned with an attempt to rebuild, to offer constructive alternatives based on the pioneering electromagnetic theory of Heinrich Hertz and on an approach, suggested by the success of the Global Positioning System (GPS), to establishing a consistent way of telling a kind of "time" divorced from environment — very like Newton's original conception. These alternatives may strike many viewers as themselves ugly — for that is an understandable response to any ringing in of the Old to replace the New.

The nice thing about science, its redeeming feature, is that human aesthetic preferences and value judgments, differences of opinion regarding ugliness and beauty, conceptions of old and new, make no lasting difference. What matters, what lasts, is what works. What works is ultimately decided by experiment. This book will not have been written in vain if it succeeds in calling attention to two particular experiments that it claims to have crucial impact. If they are done, it is just possible that what they reveal about the shape of nature will enable (and motivate) the artisans who follow to craft a better-fitting mask.

The writing of this book has been an educational experience for me. When I began it I was long aware of the shortcomings of Maxwell's equations and the superiority of Hertz's approach. But I had not conceived of the extent to which "time" needed to be reappraised, and the far-reaching consequences throughout theoretical physics of the needed reappraisal. Later, I became aware of others who preceded me in recognizing important aspects of the "alternative physics" problem. These include Charles M. Hill, Al G. Kelly, Curt Renshaw, and of course that dean of dissidents, John Paul Wesley. Unfortunately, most of these are believers in some form of fundamental substrate in our universe possessing a determinate state of motion, the equivalent of Maxwell's "luminiferous ether." In contrast, I claim reconcilability of the facts of observation with a form of relativity principle.

My book has been written with an absence of scholarly trappings such as footnotes, obligatory scientific "if's," "but's," and "on-the-other-hand's," and with a regrettable appearance of assurance on my part that I have got things right. Indeed, that is to a very limited degree my current illusion, and I have chosen my didactic mode of exposition in order to leave everywhere as little uncertainty as possible about my meaning ... but my true feelings lie closer to the quotation from Xenophanes that heads Chapter 5. In the final analysis, the book amounts to propaganda motivating a couple of simple experiments ... and the reader who knows how to get those done would save time to lay down this wordy book and at the first opportunity get busy doing them.

Colleagues and friends whose advice and moral support have eased my task include Michael H. Brill, Dennis P. Allen, Jr., Peter and Neal Graneau, J. Guala-Valverde, David F. Roscoe, Ronald G. Newburgh, C. J. Carpenter, Paul B. Coggins, Ruth R. Rains, and Kathleen Leahr. Particular recognition is due the publisher, C. Roy Keys, who—with a few others, such as Petr Beckmann, Howard C. Hayden, Harold W. Milnes, Eugene Mallove, and Cynthia K. Whitney, none of whom history (which is written by the winners) is ever likely to honor adequately—has taken a crucial leadership role among dissidents in physics and astronomy by giving them the rarest and most essential gift for any would-be contributor to science, a way to communicate. In effect these have provided the outcast, the rejected, the up-staged, the downtrodden, the politically incorrect, the wrongthinkers of physics, with a *Samizdat*. Finally, a word of special thanks to David F. Roscoe for setting aside his academic obligations to read my text and provide a much appreciated Foreword.

T. E. Phipps, Jr.
Urbana, Illinois
August, 2006

Special notice regarding length invariance: The assumption or postulate of length invariance is built into all aspects of the alternative theory developed in this book. Therefore if the reader is inherently intolerant of this idea—that is, if his mind is permanently closed against it, whether through the revelations of inner voices or through private knowledge of incontrovertible contrary experimental evidence—he should read no farther. To go on would be a waste of time, as this book is for him the purest form of crackpot literature. Those who recognize that the Lorentz contraction could be a myth are encouraged to go on. And if there are some on the edge, curious as to why anyone in the twenty-first century would propose such a bizarre idea as length invariance, full understanding requires reading the text, but a shortcut can be found in Chapter 6, Section 7.

In view of all our present difficulties it would seem that we ought at least to try to start over again from the beginning and devise concepts ... which come closer to physical reality ... If we are ever successful in carrying through such a modified treatment, it is evident that not only will the structure of most of our physics be altered, but in particular the formal approach to those phenomena now treated by relativity theory must be changed, and therefore the appearance of the entire theory altered. I believe that it is a very serious question whether we shall not ultimately see such a change, and whether Einstein's whole formal structure is not a more or less temporary affair.

—P. W. Bridgman, *The Logic of Modern Physics*

Chapter 1

What's Wrong with Maxwell's Equations?

1.1 Problems of first-order description

Virtually the whole of "established" modern fundamental theoretical physics (quantum mechanics aside) is based upon two sacred cows, Einstein's special relativity theory (SRT) and Maxwell's equations of electromagnetism, the latter being postulationally supplemented by a Lorentz force law. Of these two, Maxwell's equations are clearly the more fundamental, in that they came first chronologically—thus forming a formally prerequisite basis for SRT—and also in that they constitute the original prototype of field theory, the basis for all later "elementary particle" physics developments acceptable to modern authorities. Field theory has taken theoretical physics by storm. For most of the past century theoretical physicists have been convinced that field theory (the continuum mode of description) is the "natural language" of physics on both large and small physical scales. (I happen to disagree profoundly, but that is another story I shall not even try to tell in the present book.) For this reason it seems particularly important to get field theory right on its own terms ... a process that can be begun only by doing the same for Maxwell's equations themselves.

That will be our task in Chapters 2 and 3. In this chapter we briefly prepare the way by noting a few of the manifold shortcomings of Maxwell's equations in their accepted form (and, by incidental implication, of SRT). Before anything else, let us set down those equations for the simplest case of free space, which suffices for present purposes:

$$\vec{\nabla} \times \vec{B} - \frac{1}{c}\frac{\partial \vec{E}}{\partial t} = \frac{4\pi}{c}\vec{j}_s \qquad (1.1a)$$

$$\vec{\nabla} \times \vec{E} = -\frac{1}{c}\frac{\partial \vec{B}}{\partial t} \qquad (1.1b)$$

$$\vec{\nabla} \cdot \vec{B} = 0 \qquad (1.1c)$$

$$\vec{\nabla} \cdot \vec{E} = 4\pi\rho, \qquad (1.1d)$$

ρ and $\vec{j}_s$ being the Maxwellian charge and current densities, respectively. These four field equations, together with suitable physical boundary conditions, define and determine the electric field $\vec{E}$ and magnetic field $\vec{B}$ at the "field point" in space and time specified by the arguments (x,y,z,t) of those two vector field quantities. {I use Gaussian units (a) for the sake of history, (b) to declare my independence from *dicta* of international committees, (c) to commemorate the brightest feature of my personal higher education, that I did not go to MIT. Those wishing to use other units can consult Table 2, p. 618 of Jackson[1.1] or equivalent.}

On the broadest plane of philosophical generality, the field continuum idea is subject to the type of objection applicable to all mathematical idealizations parading as physical descriptions: The mathematical continuum, like the mathematical point, is something dwelling in the head of the mathematician. If it happens to find occasional usefulness outside that *locus*, this is more plausibly viewed as a happenstance, a bit of good luck not to be pressed too hard, than as a basis for religious experience. Yet we find in history the temporarily successful idealization, put forward tentatively by one generation, promoted into the next generation's eternal truth ... and the third generation so worshipful of it that consensus greets further experimental testing, if unsuccessful, with suspicion suited to the Anti-Science. The real sin in

all this is not in over-estimating the descriptive scope of an idealization; that is mere rashness or bad judgment. Rather, it is in
the arrogance of self-assurance-*via*-mutual-assurance through
which crowds of scholars commit cumulative, massive follies that
would be tolerated by no single scholar under restraints of individual responsibility or personal prudence. The scientific mistakes of individuals are healed and forgiven by history; the scientific arrogance of academic mobs is not. That is my comment on
today's final arbiter of scientific tastes, "consensus." But if the
greatest crime of modern physics against humanity were a crime
against humility, we could deal with that in a paragraph and be
left with no material for a book. No, it is in the details (where
God is said to reside) that we shall discover more pressing problems with existing field theory. So, nothing more will be said here
against the continuum as a descriptive abstraction.

The first preliminary to be noted about the above or any
other field continuum equations is that they offer no hint of that
fundamental aspect of nature known as the wave-particle dualism. For that, of course, Maxwell is not to blame. It is strictly in
hindsight that one recognizes all "fields" to be classical surro-
gates for spatially extended states (the wave aspect) of what
quantum theory terms *virtual particles*. These can manifest themselves locally (the observable particle aspect) only through "process completions" enabled by interaction with material objects
categorized by physicists (with typical professional parochialism)
as "detectors." *For a field to make an observable appearance, there
must always be a detector present.* From this more modern perspec-
tive, it seems that the detector is central to all observability, physics being the description of what is observable.

In that sense Maxwell's equations are not about physics, because they concern what is supposed to be abstractly true at a
disembodied mathematical "field point," not what happens cir-
cumstantially at a detector—the Maxwell field being commonly
conceived as a separately "existing" *Ding an sich* (whenever it
isn't claimed to be a "physical vector," whatever that may be, besides an oxymoron). Only by imagining the detector to occupy
the field point can the two views be reconciled ... but that exacts
a serious penalty, to be appraised presently, in that the physical
degrees of freedom of any actual (or idealized point) detector are
suppressed through such stipulated superposition upon a mathematical "field point," which by definition is entirely lacking in

motional freedoms—to the lasting and significant detriment (as it happens) of the scope of all field theory.

The most prominent deficiency to be noted about the above specific field equations is that they are not invariant under first-order (Galilean) inertial transformations. This is an extremely serious matter. It implies that in electromagnetism there exists an order of physical description—indeed, a dominantly important order, the first—at which the relativity principle (valid since Newton) does not hold. That is, if we are limited in our accuracy of physical measurement in such a way that we can observe only effects of first order in velocity, we ought seemingly to be able to *observe* actual violations of the relativity principle … such as first-order fringe shifts when our inertial system moves with respect to some "fundamental" system. This putative non-invariance can be seen from the fact that an operator such as $\partial/\partial t$, appearing in Eq. (1.1), is non-invariant under the Galilean transformation. The latter asserts that

$$\vec{r}' = \vec{r} - \vec{v}t, \quad t' = t \qquad (1.2)$$

for a first-order inertial transformation between primed and unprimed systems. (That is, $\vec{r}', t'$ and $\vec{r}, t$ specify coordinates of the same event point in the two uniformly-moving "inertial" frames, $\vec{v}$ being the constant velocity of the primed frame with respect to the unprimed one, and "first order" implying that considerations of order v^2 or higher are ignored.) From this Galilean transformation we find[1.2] the operator relations

$$\vec{\nabla}' = \vec{\nabla}, \quad \frac{\partial}{\partial t'} = \frac{\partial}{\partial t} + \vec{v}\cdot\vec{\nabla}. \qquad (1.3a,b)$$

The first of these may be derived by applying (1.2), remembering that $\vec{v}$ is constant and recognizing that in partial differential operator actions upon field quantities [traditionally considered functions of (x, y, z, t)] the chain rule applies:

$$\frac{\partial}{\partial x} = \frac{\partial x'}{\partial x}\frac{\partial}{\partial x'} + \frac{\partial y'}{\partial x}\frac{\partial}{\partial y'} + \frac{\partial z'}{\partial x}\frac{\partial}{\partial z'} + \frac{\partial t'}{\partial x}\frac{\partial}{\partial t'} = \frac{\partial}{\partial x'}, \ etc. \ \to \vec{\nabla}' = \vec{\nabla}.$$

Similarly, since $x' = x - v_x t$, $(\partial x'/\partial t) = -v_x$, etc., we have

$$\frac{\partial}{\partial t} = \frac{\partial x'}{\partial t}\frac{\partial}{\partial x'} + \frac{\partial y'}{\partial t}\frac{\partial}{\partial y'} + \frac{\partial z'}{\partial t}\frac{\partial}{\partial z'} + \frac{\partial t'}{\partial t}\frac{\partial}{\partial t'}$$

$$= \frac{\partial}{\partial t'} - \left(v_x\frac{\partial}{\partial x'} + v_y\frac{\partial}{\partial y'} + v_z\frac{\partial}{\partial z'}\right) = \frac{\partial}{\partial t'} - \vec{v}\cdot\vec{\nabla}' = \frac{\partial}{\partial t'} - \vec{v}\cdot\vec{\nabla}.$$

(Essential to this derivation is the constancy of $\bar{v}$ implied by the stipulation of inertiality.) So, although the space derivative operator is invariant under first-order inertial transformations, the partial time derivative operator is non-invariant $\left(\partial/\partial t' \neq \partial/\partial t\right)$ — a feature that prevents invariance of the field equations. Maxwell himself (who, by the way, never wrote nor saw "Maxwell's equations") was not disturbed by such non-invariance, since he did not subscribe to motional relativity and thought of electromagnetic description as subject to simplification in a fundamental system of ether "at rest."

In the nineteenth century this feature of non-invariance was taken seriously. Maxwell's predicted fringe shifts were looked for experimentally and not found. Relativity at first order was thus discovered (by Mascart and others) to be an empirical fact. That forced the conclusion that Maxwell's equations were wrong, or that something else was wrong. A "solution" was offered by Lorentz and subsequently reinforced by Einstein (in 1905). This was that "inertial" motions are to be described not by the Galilean transformation, Eq. (1.2), but by a more complicated set of equations known as "Lorentz transformations." These introduced second-order fiddlings with both space and time variables (with "Lorentz covariance" substituted for invariance) that allowed the retention of Maxwell's equations unscathed. This makes eminently good sense provided one is convinced *a priori* that Maxwell's equations are the be-all and end-all of electromagnetic physics — or that wishing and diddling can make it so. Such a conviction stands permanently in the way of any progressive evolution in the foundations of field theory and has petrified itself into a doctrine of *universal covariance*. Over the years the latter has become to theoretical physics what Virgin Birth is to Christianity.

How legitimate is it to treat first-order invariance problems by second-order solutions? In my (idiosyncratic) opinion, not at all. The Lorentz transformations for motion parallel to the x-axis are

$$x' = \frac{x - vt}{\sqrt{1 - v^2/c^2}} \tag{1.4a}$$

$$y' = y \tag{1.4b}$$

$$z' = z \tag{1.4c}$$

$$t' = \frac{t - \frac{v}{c^2}x}{\sqrt{1 - v^2/c^2}} \, . \tag{1.4d}$$

It will be observed that in (1.4d) there is a first-order term in v that must survive even if all terms in v^2 are discarded. Thus at first order these transformations take the form

$$x' = x - vt \tag{1.5a}$$

$$y' = y \tag{1.5b}$$

$$z' = z \tag{1.5c}$$

$$t' = t - \left(v/c^2\right)x \, . \tag{1.5d}$$

Evidently at sufficiently long distances (large enough x-values) there will be a first-order departure (1.5d) from the Galilean transformation (1.2), affecting the time coordinate. (In this connection, see Appendix.) Consequently on a sufficiently large scale of distance there is no limiting conformity between the Lorentz and Galilean transformations. "Inertiality"—a physical property that both common sense and Einstein tell us should not depend on choice of distance scale—is not described by the Lorentz and Galilean transformations in a mutually consistent way at first order. To believe that at first-order the Lorentz way is right and the Galilean way is wrong requires us to accept a never-verified proposition—that at great distances time and synchronized clocks behave in a different way from their counterparts in our near vicinity. This in turn means that Newtonian physics is not right on a large physical scale, even as a first-order approximation.

The reader can believe this proposition as an act of faith, if he wants to; but there is no objective basis for belief. Nor will he find it acknowledged up front in any of the SRT texts—which prefer to represent Einstein's mechanics as a covering theory of Newton's mechanics, in order to claim all the credit of the latter, preparatory to collecting extra dividends. (A "covering theory" is one that replicates all predictions of its covered theory in some parametric limiting case, but makes different predictions when that case is not realized.) Because of failure to make a first-order connection with the Galilean transformation, *no Lorentz covariant mechanical theory can be a covering theory of Newtonian mechanics.* The traditional method by which higher accuracy is attained in physics—namely, through orderly progression from lower to higher orders of approximation, each successive order being so contrived as to be a covering theory of the lower orders—is certainly violated by such an expression of faith. Let that belief (that on a large scale the physics of the first order is different from what it is locally) be yours, dear relativist, not mine. I choose to view this as a signal, a clue, that all is not well in the foundations.

In fact I cannot make it fit very well with ordinary formulations of the relativity principle, much less with assumptions of cosmological homogeneity rife among today's general relativists. The relativity principle, as normally phrased, says that the laws of nature are the same in different inertial systems. It says nothing about their being different even within one inertial system. An effect of distance on clock settings in a given inertial system might be thought not to be the same as a change of "laws of nature." Yet any first-order changing of clock settings as a function of distance, departing from the Galilean $t' = t$, will certainly affect at first order that law of nature known as Newton's second law of motion, $\vec{F} = m\vec{a}$, which is founded unequivocally on $t' = t$.

On the other hand, if we accept that *locally* the Lorentz transformations reduce at first order to the Galilean ones, then the "relativistic" theory seemingly inherits at first order the difficulty of Maxwell's predicted first-order fringe shifts under the Galilean transformation, discussed above—the shift-disproving experiments being of *local* character and at first order. As it happens, there is a save based on *Potier's principle* (a deduction from Fermat's principle), whereby the putative fringe shifts prove to be theoretically unobservable after all. This will be one of our topics in Chapter 2. The matter is a bit subtle, and is never discussed nor alluded to in modern texts of electrodynamics. This is part of the general conspiracy of silence concerning first-order electromagnetic physics.

As nearly as I can tell, the claim that at first order a physical inertial transformation is described on the timelike side by $t' = t - (v/c^2)x$ is without a shred of observational support. Rather the contrary would seem to be the case: Since the Earth reverses its velocity every six months, it switches inertial systems with the corresponding frequency, so all its co-moving clocks throughout space should require continual resetting in those ever-changing states of motion, in such a manner that the apparent timings of astronomical events at great distances (large x-values) should vary with an annual period, and should vary (from correspond-

ing event timings in our immediate spatial vicinity) proportionally to x. The first-order annual velocity change effect resulting
from Eq. (1.5d) would thus imply a first-order annual change of
clock readings and apparent dynamical evolutions at the locations of distant galaxies. There is no way clock readings (settings)
at a given place can change periodically without associated clock
rates appearing to change (as an artifact of the continual clock re-
settings associated with frame changes, about which the reader
can learn from standard SRT texts such as Taylor and
Wheeler[1.3]—see the twin paradox discussion there).

In other words the earth's circling should periodically affect
the *apparent* (measured) rates of remotely distant physical processes—although the processes themselves do not factually vary in
rate. But in actuality no such distance-dependent appearances
appear. Note that at first order there is no space-side fiddle (contraction) to "compensate" the claimed time-side fiddle (clock
phase changes)—as there is at second order—since the space coordinates at first order transform in Galilean-Euclidean fashion,
Eqs. (1.5a-c). The Lorentzian "first-order world" defined by Eq.
(1.5) thus seems decidedly unworldly.

If our argument about distant astronomical process appearances is correct, it would seem that SRT and its mathematics of
"Lorentz covariance" (too often sloppily referred to as "Lorentz
invariance" or, heaven help us, simply "invariance") get by on the
strength of an intergalactic gentlemen's agreement to ignore the
observational physics of the first order. This is such an obvious
and trivial criticism that it is a waste of time to make it, since
minds are closed. Dingle,[1.4] who was put through the wringer by
the physics establishment and accused of "dementia" for his alleged crime of failing to "understand" SRT, remarked, "It is ironical that, in the very field in which Science has claimed superiority
to Theology, for example—in the abandoning of dogma and the
granting of absolute freedom to criticism—the positions are now
reversed. Science will not tolerate criticism of special relativity,
while Theology talks freely about the death of God, religionless
Christianity, and so on." He found out the hard way that *when all
gentlemen subscribe to an agreement, all non-subscribers are by definition non-gentlemen.* It is to this pass that SRT has brought the
physics of the twenty-first century. And there, for all I can tell, it
will stick—perhaps for millennia. One can only hope that a single
millennium will suffice to eradicate the Einsteinian folly, as it did

the Ptolemaic. But the durability of a myth grows nonlinearly with its perceived beauty—which is far more a matter of self-consistency than of external-world-consistency.

1.2 The under-parameterization of Maxwell's equations

Much lip service is paid by Maxwell's followers to "source-sink reciprocity." Oddly, however, in the equations themselves there is no reciprocity of *parameterization* between source and sink motions. Look closely at Maxwell's equations, (1.1). You will see that source motions are parameterized by $\vec{j}_s = \rho \vec{v}_s$, where $\vec{v}_s$ is indeed the velocity in the observer's inertial frame of the particle that is the source of the field (or a corresponding charge density). But look as you may, you will find no parameter describing velocity of the field detector, absorber, or sink. Now, pause and reflect … What is going on? After all, the sink is physically as important to the radiation process as the source (no photon being able to land without a landing-place), and far more so for purposes of observation or "measurement"—recognized in quantum theory as the quintessence of any micro process. In field theory what has become of this quintessence?

What is going on is that the detector must be pictured as permanently at rest at the field point. Remember the observer's "field point"? Well, that's where the sink is. And nowhere else. Ever. Think of that! In this fickle world of motion, flux, and relativity, here is something eternal, fixed, immobile … something you can count on. Dare I say it … ? Something absolute. The source, in contrast, is not absolute. Like everything else in nature it can move freely, knows no home, has its velocity and degrees of freedom parameterized, *etc.* But the sink by definition sticks like glue to "the observer," as a sort of extension of his personality. Since the observer is always by definition inertial, so is the sink. Here is a composition of matter (try making a detector without matter) that is by definition always inertial in its motions. Strive as it may to break that bond with the inertial observer, the poor thing just cannot do so consistently with Maxwell's field theory. *The sink lacks parameters that would grant it **in theory** those physical degrees of freedom which it clearly possesses **in nature**.* It is always dangerous to build theory that departs from nature in ways so elementary you can count them on one hand, even with a thumb and finger missing.

Looking at it another way, we can view the observer as tied to a particular composition of matter (the field detector). He thus becomes a "preferred observer" with respect to that bit of matter. The state of motion of the matter in question then defines a fundamental physical inertial system that is preferred (by that observer) above all others. Clearly, in either way of looking at it, an element of absolutism contrary to the relativity idea has entered in a basic way—an element ineradicable as long as we refuse (by declining to parameterize sink motions) to allow arbitrary relative motions between every observer and every sink. The motional relativity *idea*, in contrast, properly requires that all observers be free to move with respect to all compositions of matter. There must be no preferred observer or bit of matter, ever, anywhere in the universe. Pause for a moment to reflect on the present claim: Here in the heart of Maxwell's equations is a completely anti-relativistic structural element—an omission of parameterization that prevents relative motion between the observer and a corresponding favored composition of matter. And Maxwell's equations lie at the heart (formally and structurally) of SRT, which is billed as the purest expression of the relativity idea. In other words, in the heart of the heart of "relativity," as we know it today, dwells a heartworm.

1.3 The problem about Faraday's observations: d/dt

A directly related difficulty evidenced by the Maxwell magnetic induction equation, (1.1b), is that it misrepresents the Faraday observations on which it is allegedly based. According to all the texts, Faraday's observations are described by

$$\oint \vec{E} \cdot d\vec{\ell} = -\frac{d\Phi}{dt} = -\frac{d}{dt} \iint \vec{B} \cdot d\vec{S}, \qquad (1.6)$$

where Φ is the $\vec{B}$-field flux through the surface bounded by a closed electrical circuit. Note particularly the appearance of the total time derivative d/dt here. The line integral on the left represents the electromotive force (*emf*) generated in the flux-penetrated circuit when any of various experimental parameters are changed. Among those changed by Faraday was the *shape* of his circuit. That is, he moved *part* of the circuit in a magnetic field and observed that this produced an *emf*. It is this shape-changing aspect that necessitates using a total time derivative operator d/dt in (1.6), rather than the partial derivative $\partial/\partial t$ native to traditional field theory. There is no escape from d/dt, because a shape change cannot occur without accelerated relative motions of circuit parts. Such different motions in different places require for their *localized* (differential) description different values of a local velocity parameter $\vec{v}_d(t)$, of the sort that is present in d/dt but not in $\partial/\partial t$. There being no inertial system in which the circuit as a whole can be considered even approximately "at rest," there is no applicability of SRT, which by its basic terms of reference is limited to inertial motions.

According to the ideology of relativists, such problems of mixed local accelerations and non-accelerations should rigorously be handled by general relativity theory (GRT). As far as I know, the doctrine to that effect has never been put into practice in this case, perhaps because even relativists—though a rigorously sober lot, as befits those teetering on the very tippy-top rung of the intellectual ladder—have a sufficient sense of humor to smile at any claim of logical *necessity* to treat Faraday's elementary observations *via* four-index tensor symbols (or to drag in *equivalence* of acceleration to "gravity"), when a simple total time derivative will manifestly do the job with elegance and precision. So, let us not argue with the electrodynamics texts, but take their word for (1.6) as a statement of empirical fact. How, then, do relativists and field theorists get from (1.6), with its empirically correct d/dt, to Maxwell's (1.1b), with its politically correct $\partial/\partial t$? Ah, now that is a tale that would interest psychologists. Some authorities, such as Panofsky and Phillips[1.5] just brutally set down a howler,

$$\oint \vec{E} \cdot d\vec{\ell} = -\frac{d}{dt} \iint \vec{B} \cdot d\vec{S} = -\iint \frac{\partial \vec{B}}{\partial t} \cdot d\vec{S} . \qquad (?)$$

If you can square that with your mathematical conscience you are off to the races in "deriving" the Maxwell (1.1b), involving a partial time derivative. These authors stifle their scruples by fast talk about "a differential expression valid for free space or a stationary medium." Indeed, this might be a legitimate representation of Faraday's observations if they had been confined to free space or a stationary medium. However, they involved changing the shape of a circuit—altering the path of $d\vec{\ell}$ in flux-penetrated space. That's a different matter, which mathematically requires retention of the d/dt operator.

Or, take an alternative dodge perpetrated by J.D. Jackson[1.1] and reflecting equivalent desperation to get to the known right answer. He says, "Faraday's law can be put in differential form by use of Stokes's theorem, provided the circuit is held fixed in the chosen reference frame … " So, his draconian choice is simply to ignore the experiment and pretend that Faraday held his circuit fixed in an inertial frame. That gets us to Maxwell's (1.1b), all right … but at what a cost! Lorrain and Corson[1.6] carry out a variant of this reasoning even more conscientiously. If I understand their analysis, they allow non-inertial motions, but require the circuit at all times to move as a rigid whole—still in deliberate disregard of what Faraday did and saw. Wangsness[1.7] claims to allow shape changes of the Faraday circuit, but when his formulas are examined they are found to contain only a single velocity-dimensioned parameter $\bar{v}$, our old SRT friend, the relative velocity of two inertial systems … so, although the shape change was talked about, by what was it parameterized? When you change the shape of a circuit you impart different velocities to different *portions* of it. If your theory lacks a velocity parameter that can take different numerical values on different circuit portions, your mathematics is plainly inadequate to the physics … though it may be adequate to the political demands of your day.

From the variety of such swindles by modern authorities—all of whom insist on making bricks without straw (reducing global circuit descriptions to differential form without benefit of local shape-change parameterization)—and the unanimity of critical silence by which they are greeted, the degree of degeneracy of modern physical judgment and ethics could be judged … and mathematical integrity as well … if anyone were judging. But physics has shown itself a *profession* (like others) mightily resistant to whistle-blowing.

Some more sentient observers of the electromagnetic scene have noticed that *information is lost* in passing from the integral formulation (1.6) to the differential formulation (1.1b). Consequently they insist on integral formulations of field theory as more "fundamental." [Thus (1.6) allows for continuous deformations of the global integration contour, whereas (1.1b) contains no way of describing the resulting continuous local departures from inertial motion.] None of these experts has recognized that the simple way to recover *all* the lost information is by *changing the differential formulation* to make it include an extra velocity parameter $\vec{v}_d(t)$ capable of describing the said local departures by taking on different local values ... and the simple way to do that is to avoid the strenuous mental gymnastics involved in replacing d/dt by $\partial/\partial t$, and instead to relax and peacefully leave d/dt in the differential formulation. This is the sort of thing readily grasped by freshmen but utterly hidden from the ken of academic savants (unless their peers are telling them about it, in which case they invented it back in '01).

1.4 Justification for a Hertzian form of Faraday's law

Why this concerted professional willingness or compulsion to make obvious elementary mathematical mistakes in the interest of getting to Eq. (1.1b), as if at all costs? Simply because the stakes are tremendous. The costs are indeed "all." The reputation of every physicist of the modern era, dead or alive, depends on that little $\partial/\partial t$. Empiricism calls for the field equation (1.1b) to be replaced by the Hertzian invariant form

$$\vec{\nabla} \times \vec{E} = -\frac{1}{c}\frac{d\vec{B}}{dt}, \qquad (1.7)$$

but that would destroy *spacetime symmetry*, which arises from and depends critically upon the appearance of the time variable t in the field equations (1.1) of electromagnetism solely in a balanced form

$$\left(\frac{\partial}{\partial x}, \frac{\partial}{\partial y}, \frac{\partial}{\partial z}, \frac{\partial}{\partial ict} \right)$$

that manifests a fundamental *symmetry* of space and time partial derivative operators. This is the mother lode—where spacetime symmetry originates. The total time derivative, in contrast, is traditionally expressed as

$$\frac{d}{dt} = \frac{\partial}{\partial t} + \left(\vec{v}_d \cdot \vec{\nabla} \right), \qquad (1.8)$$

where $\vec{v}_d$ may be treated as constant or more generally as an arbitrary function of t (at any rate constant under the action of $\vec{\nabla}$). To use d/dt is to upset the balance of space and time:

$$\left(\frac{\partial}{\partial x}, \frac{\partial}{\partial y}, \frac{\partial}{\partial z}, \frac{\partial}{\partial ict} + \left(\vec{v}_d \cdot \vec{\nabla} \right) \right).$$

So, it's hello d/dt, goodbye spacetime symmetry. Note that this $\vec{v}_d$ is a local velocity parameter entirely different from the global parameter $\vec{v}$ descriptive of an inertial frame transformation. The proof of Eq. (1.8) for the general case in which the field detector is treated as a point particle in arbitrary (possibly accelerated) motion, $\vec{v}_d = \vec{v}_d(t)$, is immediate from the chain rule:

$$\frac{d}{dt} = \frac{dx}{dt}\frac{\partial}{\partial x} + \frac{dy}{dt}\frac{\partial}{\partial y} + \frac{dz}{dt}\frac{\partial}{\partial z} + \frac{dt}{dt}\frac{\partial}{\partial t} = \frac{\partial}{\partial t} + \vec{v}_d \cdot \vec{\nabla}. \tag{1.9}$$

It is important to note, as an implicit feature of this derivation and conception, that the field quantities or other operands of d/dt are not viewed as arbitrary functions of (x,y,z,t), as is standard in traditional field theory, but are instead viewed as functions of $(x(t),y(t),z(t),t)$. This means that one maintains always an inflexible regard for the fact that all field values are measured quantities—measured *on the trajectory* of the point-like detector ... that trajectory being described by $\vec{r}(t) = (x(t),y(t),z(t))$ at any *locus* in space where the "field" is to be interrogated, or at each place along a conducting circuit subject to deformations. Consequently, our $\vec{v}_d$ is *not* a "velocity field."

The introduction of the d/dt operator completely spoils the formal symmetry of space and time differentiations and thus destroys the basis (in electromagnetism) for SRT and for all modern physics built upon it. And it leaves no justification for "universal covariance," the mathematical expression of spacetime symmetry that is the touchstone or shibboleth of our scientific age. The appearance here of an extra velocity-dimensioned parameter $\vec{v}_d$ fits hand-in-glove precisely to compensate the Maxwellian under-parameterization mentioned in Section 1.2. Exploitation of this lucky fit will be our main objective in the next chapter.

The upshot of the present discussion is that the physicist faces a choice between respecting empirical fact, as embodied in Faraday's observations (demanding total time differentiation), and imposing mathematical beauty, as embodied in covariant formalisms (for which it is essential that all time differentiation operators be of the partial type). Among theorists of our era there has never been a moment's hesitation about that choice. They have stopped at nothing to legislate their foreknown truth. This is to say that the people who currently call themselves, and are paid to be, *physicists* are flying under false colors. They are ideologues, playthings of their collective infatuation with beauty. The world as it is interests them less than the world as it ought to be. They are what happens when physicists morph into mathematicians (of sorts). But since they constitute an intellectual monopoly that controls all media of communication in their field, whistle-blowing by individual out-groupies like myself is a vain endeavor. Those who control today's scientific communication exert absolute power to allow or prevent progress, and by that power have been corrupted absolutely.

1.5 Other problems of Maxwell's equations

In conclusion regarding the deficiencies of Maxwell's equations, we have barely scratched the surface of this bountiful topic. Plenty more such are to be seen by any beholder not blinded by science. A major portion of this book will, in one way or another, be concerned with exploring the ramifications of this rich subject. I have thus far omitted criticism, for example, of sins against logical economy, such as that entailed in defining a "field" as *force on unit charge* and then postulating a separate *force law for charges*. And I have omitted criticism of internally contradictory aspects of field theory itself. For instance, Eq. (1.6) defines "flux" as an integral. This implies that any circuit senses *instantly via* its *emf*— *e.g.*, by a set of voltmeters placed everywhere around the perhaps infinitely spatially extended circuit—any change of a global (integral) property. (If not, please tell me which voltmeter measures the flux change first. And feel free to place yourself in any inertial system!) This can only betoken instant and simultaneous actions at-a-distance—supposedly forbidden by the very terms of reference of field theory, not to mention SRT. Yet here we perceive field theory (always presented as the bastion of causal thinking) to be founded on a conception that is manifestly acausal and contrary to the relativity of simultaneity. The bones of quantum mechanics move perceptibly beneath the skin of the Maxwell field[1.8]—instant action-at-a-distance being an integral aspect of quantum theory. The only known exception in the entire range of physical experience to the rule of instant action is radiation (locally completed quantum processes)—the tail that hitherto has wagged the dog.

Suffice it to say that Maxwell was a genius. He earned that status by eclipsing the original geniuses of electrodynamics, André-Marie Ampère and Wilhelm Weber. It is the fate of genius to be eclipsed. *Pace* ... let genius be sufficient unto the day thereof. The whole concept of the field-continuum mode of description is over-ripe for reappraisal, if our era could move beyond its complacently dismissive view of electrodynamics as a closed book. About this subject there is only one thing *trivial*, in the sense of being true clearly to any child—namely, that it is barely (and thus far poorly) begun.

1.6 Chapter summary

In this chapter we have concentrated on three major deficiencies of Maxwell's equations:

- They are not invariant at the lowest (first) order of approximation under first-order inertial (Galilean) transformations.
- They are under-parameterized, in that source motions are described but not sink motions. Such an implicit promotion of the sink to a preferred motional status flouts any conceivable form of relativity principle. (This is why invariance fails.)
- They misrepresent Faraday's observations of induction through the false implication that $\partial/\partial t$ can treat that for which d/dt is mathematically necessary: *viz.*, description of the *emf* generated by *changing the shape* of a conducting circuit penetrated by magnetic flux.

Other shortcomings are readily cited, but these will suffice to give us the clues needed in order to commence a plausible reconstruction of electromagnetic theory in the next chapter.

References for Chapter 1

[1.1]   J. D. Jackson, *Classical Electrodynamics* (Wiley, New York, 1962).

[1.2]   M. Jammer and J. Stachel, *Am. J. Phys.* **48**, 5 (1980).

[1.3]   E. F. Taylor and J. A. Wheeler, *Spacetime Physics* (Freeman, San Francisco, 1966).

[1.4]   H. Dingle, *Science at the Crossroads* (Martin, Brian, and O'Keefe, London, 1972).

[1.5]   W. K. H. Panofsky and M. Phillips, *Classical Electricity and Magnetism* (Addison-Wesley, Reading, MA, 1962), 2nd ed.

[1.6]   P. Lorrain, D. R. Corson, and F. Lorrain, *Electromagnetic Fields and Waves: Including Electric Circuits* (Freeman, New York, 1988), 3rd ed.

[1.7]   R. K. Wangsness, *Electromagnetic Fields* (Wiley, New York, 1979).

[1.8]   C. A. Mead, *Collective Electrodynamics* (MIT, Cambridge, MA, 2001).

> There is no physical phenomenon whatever by which light may be detected apart from the phenomena of the source and the sink ... Hence from the point of view of operations it is meaningless or trivial to ascribe physical reality to light in intermediate space, and light as a thing travelling must be recognized to be a pure invention.
>
> —P. W. Bridgman, *The Logic of Modern Physics*

Chapter 2

What to Do About It ... (the Hertzian Alternative)

2.1 First-order invariant field equations

The criticisms of Maxwell's field equations in the preceding chapter were made in order to gain constructive clues to the building of alternative electrodynamic theory. In that we have been successful: We saw that those equations were *under-parameterized* (lacking a velocity-dimensioned parameter to describe sink motions), and also that they misrepresented Faraday's observations of induction by using the non-invariant operator $\partial/\partial t$ instead of the first-order *invariant* operator d/dt (which contains an extra velocity-dimensioned parameter $\vec{v}_d$ needed to describe local departures of his circuit from inertiality). Putting these clues together—matching parameter deficit with parameter surplus—we have the inference forced upon us that it would be worthwhile to aim for an invariant (instead of covariant) formulation of electromagnetism in terms of total time derivatives, wherein $\vec{v}_d$ is interpreted as describing *sink motion*. (As a confirmation and dividend for using an invariant time derivative operator, we shall discover that the resulting field equations are themselves invariant at first order.) No inspiration can be claimed for such an idea, or confluence of ideas, since the facts make it inescapable. The idea, over-ripe, has fallen from the tree and has only to be picked up. We know that motional relativity is an empirical fact, and that invariance is the natural mathematical expression of a relativity principle. Covariance is supposed to do as well, but certainly not better! So, first things being first, let us get busy and get first-order electromagnetic field theory right for a change. That will be our aim in this chapter. Then, in the next chapter, we shall see whether higher-order description can be induced to follow suit (as a covering theory) in the course of nature … as nature, so to speak, intended.

To prepare the way, we gather up a loose end by proving our assertion of the *invariance* of d/dt under the Galilean (first-order) inertial transformation, Eq. (1.2). Applying the Galilean velocity addition law,

$$\vec{v}_d' = \vec{v}_d - \vec{v},$$

(2.1)

where $\vec{v}_d$ is sink or detector velocity measured in the unprimed inertial frame, $\vec{v}_d'$ is the same measured in the primed frame, and $\vec{v}$ is velocity of the primed relative to the unprimed inertial frame, together with Eqs. (1.3) and (1.8), we find

$$\left(\frac{d}{dt}\right)' = \left(\frac{\partial}{\partial t} + \vec{v}_d \cdot \vec{\nabla}\right)' = \frac{\partial}{\partial t'} + \vec{v}_d' \cdot \vec{\nabla}'$$

$$= \left(\frac{\partial}{\partial t} + \vec{v} \cdot \vec{\nabla}\right) + \left(\vec{v}_d - \vec{v}\right) \cdot \vec{\nabla} = \frac{\partial}{\partial t} + \vec{v}_d \cdot \vec{\nabla} = \left(\frac{d}{dt}\right),$$

(2.2)

which verifies the first-order invariance of d/dt. Note that in this proof it is essential to distinguish between the two velocity-dimensioned parameters, $\vec{v}$ and $\vec{v}_d$. They are physically and mathematically unrelated, $\vec{v}$ being always necessarily constant for inertial motions, and $\vec{v}_d$ being more generally descriptive of arbitrary non-inertial motions *via* an arbitrary Lagrangian particle trajectory, $\vec{v}_d = \vec{v}_d(t)$ —but not *via* an Eulerian "velocity field" $\vec{v}_d = \vec{v}_d(x,y,z,t)$. The latter requires a more elaborate analysis, patterned on that of Helmholtz,[2.1] who, according to Miller,[2.2] derived the result

$$\frac{d\vec{U}}{dt} = \frac{\partial \vec{U}}{\partial t} + \vec{V}\left(\vec{\nabla} \cdot \vec{U}\right) - \nabla \times \left(\vec{V} \times \vec{U}\right).$$

If we apply this to arbitrary vector fields, $\vec{U}, \vec{V}$, then a standard vector identity[2.3] converts it to

$$\frac{d\vec{U}}{dt} = \frac{\partial \vec{U}}{\partial t} + \left(\vec{V} \cdot \vec{\nabla}\right)\vec{U} + \vec{U}\left(\vec{\nabla} \cdot \vec{V}\right) - \left(\vec{U} \cdot \vec{\nabla}\right)\vec{V}. \qquad (2.3)$$

If $\vec{\nabla}$ operating on $\vec{V}$ yields zero, then (2.3) reduces to (1.8). Such results have been given also by Abraham-Becker[2.4] and more recently by Dunning-Davies,[2.5] as well as by other authors. They provide a generalization of the traditional total or convective derivative (1.8) from the Lagrangian specialized case $\vec{v}_d = \vec{v}_d(t)$, under which (1.8) and (2.2) are valid, to the general field-theoretical case, $\vec{V} = \vec{V}(x,y,z,t)$. However, I see no use for such generality in the present context. Physically (setting aside entirely all those hardy perennial interpretations, of the Lorentz ether pedigree, that postulate an electromagnetic "medium" in an identifiable state of motion), they correspond to the case of an extended "mollusk" detector (to borrow a figure of speech from Einstein), or electric eel—which could be expanding and contracting, with non-vanishing $\vec{V}$-divergence or gradient. Prosaically, attention will be confined in this book to idealized inanimate *point detectors*—hence to the Lagrangian $\vec{v}_d = \vec{v}_d(t)$.

We proceed to postulate the first-order invariant electromagnetic field equations, here called Hertzian because—though often (and still continually being) re-invented—they were first proposed by Heinrich R. Hertz.[2.6] Getting rid of the non-invariant $\partial/\partial t$ everywhere in Maxwell's equations in favor of the invariant d/dt, we have for free space, to which attention will be confined here:

$$\vec{\nabla} \times \vec{B} - \frac{1}{c}\frac{d\vec{E}}{dt} = \frac{4\pi}{c}\vec{j}_m \qquad (2.4a)$$

$$\vec{\nabla} \times \vec{E} = -\frac{1}{c}\frac{d\vec{B}}{dt} \qquad (2.4b)$$

$$\vec{\nabla} \cdot \vec{B} = 0 \qquad (2.4c)$$

$$\vec{\nabla} \cdot \vec{E} = 4\pi\rho. \qquad (2.4d)$$

These equations are identical to the Maxwell Eqs. (1.1) except that d/dt replaces $\partial/\partial t$ and the source terms are interpreted differently. In (2.4a) the modified source term $\vec{j}_m$ is interpreted as the current density that is *measured* by a suitable point detector moving at velocity $\vec{v}_d(t)$ in the observer's inertial frame. It is evidently related to the Maxwellian source current density $\vec{j}_s$

(measured at the stationary field point by a detector at rest there) by the first-order vector summation,

$$\vec{j}_m = \vec{j}_s - \rho\vec{v}_d. \tag{2.5}$$

The $-\rho\vec{v}_d$ term here represents an equivalent current (of reversed direction, signified by the minus sign) due to detector motion at velocity $\vec{v}_d$. In effect, a current-density detector is pictured as passing with velocity $\vec{v}_d$ through the observer's "stationary" field point and as having a snapshot taken of it there at the moment of detection. The snapshot shows it to be displaying on its read-out a measured value $\vec{j}_m$, which is the vector sum of the Maxwell source current and the convection current produced by its own motion. It will be observed that (2.4b) reproduces the empirical result (1.7), which we decided in Chapter 1 was needed to describe Faraday's observations. And since there is no virtue in going half-way with invariance, the invariant total time derivative has been used in (2.4a) as well. By the nature of invariance, the Hertzian invariant formulation of field theory obeys a first-order relativity principle and gets rid automatically of the spurious first-order fringe shifts predicted by Maxwell's non-invariant equations when subject to the Galilean transformation.

We stipulate that various field-related quantities such as source charge and current are measured (simultaneously) at the field point, as are the field quantities $\vec{E}$ and $\vec{B}$ themselves. Moreover, the various physical detectors (as of electric and magnetic fields, charge densities, *etc.*) that make such measurements all share a common state of motion parameterized by $\vec{v}_d$. Because of this shared-motion assumption we may picture all these quantities as measured by a single multi-purpose "detector." This we idealize as a mathematical point (but one sufficiently massive to possess a classical trajectory), which moves through the observer's idealized stationary "field point" and coincides with it at the instant of measurement. This constitutes a minor extension of the (seldom-discussed) idealizations employed by Maxwell and field theorists generally—who tacitly assume immobility of the detector at the field point. The term "point detector," as used here, will always refer to a physical detector, usually of several field-related quantities, idealized as a mathematical point, yet recognizable as surrogate for an actual macro instrument through the fact of its possessing a well-defined trajectory. Such preliminaries need not be dwelt on further. They are usually skipped over almost entirely by textbook expositors, who break a leg to get to the mathematics. The more the textbook-writing professors cripple themselves in this way, the more their students do the same when they in turn grow up to be textbook-writing professors.

Since the Maxwell case of a stationary detector represents the special case $\vec{v}_d = 0$, $d/dt \rightarrow \partial/\partial t$ [as shown by Eq. (1.8)], we see that Hertz's electromagnetism, Eq. (2.4), constitutes a "covering theory" of Maxwell's equations, (1.1). That is, all predictions of Maxwell's theory are reproduced by Hertz's theory in the special case that the field detector is at rest, $\vec{v}_d = 0$, at the field point in the observer's inertial system. If the detector moves with respect to the field point $(\vec{v}_d \neq 0)$, new effects are predicted by Hertz … and the prediction is invariant at first order, so that all inertial observers agree on it. The electromagnetic "laws of nature" are the same for all of them, so a first-order relativity principle is automatically obeyed in conformity with the mathematical invariance (under Galilean transformations). Einstein or Maxwell might handle this physical situation of detector motion through the field point by transferring observer and measuring apparatus bodily into another frame, also required to be inertial … and then forever after would be limited to strictly inertial motion of a detector fixed at the observer's new field point. The Hertzian analysis, not being limited to inertial motions of the sink, allows arbitrary *non-inertial* $\vec{v}_d = \vec{v}_d(t)$. It is thus a "covering theory" in more ways than one.

Since the operators $\vec{\nabla}$ and d/dt appearing in the Hertzian field equations (2.4) are Galilean invariant [per Eqs. (1.3) and (2.2)], and the source terms ρ and $\vec{j}_m$ are measured quantities on which all observers must agree, it follows that the field equations must be Galilean invariant; consequently that the field transformation law at first order is

$$\vec{E}' = \vec{E}, \quad \vec{B}' = \vec{B}; \tag{2.6}$$

that is, the field quantities themselves are invariant. Let us pause to prove our claimed *field equation invariance* in detail. First we note the Galilean source transformation equations

$$\rho'(\vec{r}',t') = \rho(\vec{r},t) \tag{2.7}$$

and

$$\vec{j}_s'(\vec{r}',t') = \vec{j}_s(\vec{r},t) - \rho(\vec{r},t)\vec{v}, \tag{2.8}$$

this $\vec{v}$ being the velocity parameter appearing in the Galilean transformation, Eq. (1.2). Eq. (2.7) follows from the facts (a) that $\vec{r}'$ and $\vec{r}$ refer to the same field point in space, viewed in primed and unprimed inertial systems, (b) that $t' = t$, and (c) that charge density is measured by counting charges in the given small (infinitesimal) detector volume, which has the same size and shape as viewed in each system and regardless of its relative motion (because we postulate length invariance, as will be discussed more fully in due course—cf. Chapter 6, Section 7). Such a charge count, a pure number, must be invariant under changes of the viewing system. Similarly (2.8) holds for the Maxwellian current density, with inclusion of a convective current-density term due to the inertial relative motion. These are well-known results from Maxwell theory, to be found for instance in Jammer and Stachel.[1.2] From this we prove the invariance of the Hertzian measured current density,

$$\vec{j}'_m = \left(\vec{j}_s - \rho \vec{v}_d\right)' = \vec{j}'_s - \rho' \vec{v}'_d = \left(\vec{j}_s - \rho \vec{v}\right) - \rho\left(\vec{v}_d - \vec{v}\right)$$
$$= \vec{j}_s - \rho \vec{v}_d = \vec{j}_m,$$

(2.9)

where use has been made of Eqs. (2.1), (2.5), (2.7), and (2.8). With these preparations, proof of the Hertzian field equation invariances becomes a matter of inspection.

Thus invariance of (2.4c),

$$\left(\vec{\nabla} \cdot \vec{B}\right)' = \vec{\nabla}' \cdot \vec{B}' = \vec{\nabla} \cdot \vec{B} = 0,$$

(2.10)

follows from (2.6) and (1.3a). Invariance of (2.4d),

$$\left(\vec{\nabla} \cdot \vec{E} - 4\pi\rho\right)' = \vec{\nabla}' \cdot \vec{E}' - 4\pi\rho' = \vec{\nabla} \cdot \vec{E} - 4\pi\rho = 0,$$

(2.11)

follows from (2.6), (1.3a), and (2.7). That of (2.4b),

$$\left(\vec{\nabla} \times \vec{E} + \frac{1}{c}\frac{d\vec{B}}{dt}\right)' = \vec{\nabla}' \times \vec{E}' + \frac{1}{c'}\left(\frac{d}{dt}\right)' \vec{B}' = \vec{\nabla} \times \vec{E} + \frac{1}{c}\frac{d\vec{B}}{dt} = 0, \quad (2.12)$$

follows from (2.6), (1.3a), (2.2), and the assumed first-order constancy of the units ratio,

$$c' = c.$$

(2.13)

Finally, the invariance of (2.3a),

$$\left(\vec{\nabla} \times \vec{B} - \frac{1}{c}\frac{d\vec{E}}{dt} - \frac{4\pi}{c}\vec{j}_m\right)' = \vec{\nabla}' \times \vec{B}' - \frac{1}{c'}\left(\frac{d}{dt}\right)'\vec{E}' - \frac{4\pi}{c'}\vec{j}'_m$$

$$= \vec{\nabla} \times \vec{B} - \frac{1}{c}\frac{d\vec{E}}{dt} - \frac{4\pi}{c}\vec{j}_m = 0,$$

(2.14)

follows from the above together with (2.9). In the same way we may identify a first-order *invariant continuity equation*, which generalizes the customary one involving $\partial/\partial t$:

$$\left(\vec{\nabla}\cdot\vec{j}_m + \frac{d\rho}{dt}\right)' = \vec{\nabla}'\cdot\vec{j}'_m + \left(\frac{d}{dt}\right)'\rho' = \vec{\nabla}\cdot\vec{j}_m + \frac{d\rho}{dt} = 0 .$$

(2.15)

The generalization resides, of course, in the circumstance that the detector measuring total current $\vec{j}_m$ can move with respect to the field point; whereas such motion is (without physical justification) forbidden in the traditional $\partial/\partial t$ formulation. The above invariances all hold at first order under the Galilean transformation, Eq. (1.2). We shall have more to say presently about the physical meaning of a field invariance such as (2.6), which is entirely different from the corresponding "covariance."

Finally, we note in passing that Eq. (2.1), $\vec{v}'_d = \vec{v}_d - \vec{v}$, implies various expressions of the reciprocity idea. Thus, if we formally put primes on both sides of this equation we get $\vec{v}''_d = \vec{v}'_d - \vec{v}'$. Interpreting one prime as effecting a switch from one inertial frame to another, and two primes as effecting a switch back, we have the requirement $\vec{v}''_d = \vec{v}_d$, which implies $\vec{v}''_d = \vec{v}_d = \vec{v}'_d - \vec{v}'$, or $\vec{v}'_d = \vec{v}_d + \vec{v}'$. Comparing with (2.1), we see that

$$\vec{v}' = -\vec{v} ,$$

(2.16)

which is the customary expression of Galilean velocity reciprocity between inertial frames. If the latter is taken as given, the argument can be worked backwards to prove $\vec{v}''_d = \vec{v}_d$.

2.2 History: Why did Hertz fail?

It is of interest to digress for a moment to see how history explains why the reader has never heard of Hertz's first-order invariant version of electromagnetic theory. First, let it be noted that Heinrich Hertz, although generally credited only as the experimentalist who confirmed in his laboratory Maxwell's ideas about the wave propagation of light, was also a powerful theorist in his own right. In the last chapter of his book *Electric Waves*,[2.6] Hertz developed a formally (Galilean) invariant generalization of Maxwell's theory, involving a new velocity-dimensioned parameter with components (α, β, γ). He conceived of his theory (formally the same as that developed above) as describing an electrodynamics of "moving media," and interpreted his new velocity parameter as ether velocity. This was a serious mistake, a false interpretation. He compounded that error by postulating a Stokesian ether 100% convected by ponderable matter. This made his theory testable, because it reified the ether—giving it "hooks" to observable matter. (Maxwell, too, assumed an ether, but cleverly avoided giving it hooks to anything observable!) Soon after Hertz's death an experimentalist, Eichenwald,[2.7] went into his laboratory and disconfirmed Hertz's predictions. The invariant theory was thus discredited and relegated to history's trash bin.

One important lesson to be learned: It is never theory alone that is proven or disproven in the laboratory; it is theory plus interpretation of the symbols employed in it. We have seen that by the simple expedient of interpreting Hertz's (α, β, γ) as our $\vec{v}_d$, with no reference to "ether"—specifically, by re-interpreting Hertz's "ether velocity" as *field-detector velocity* relative to the laboratory inertial system—all conflicts with observation are eliminated through the obtaining of a covering theory of Maxwell's electromagnetism. In fact Hertz possessed such a *formal* covering theory, but spoiled it as physics by his bad guess about symbol interpretation. The irony of history is that Maxwell's (ether) interpretation, too, was discarded, leaving to posterity only his formalism. That residual (once and future) formalism was inferior to the discarded Hertzian formalism, in exactly the sense that any non-invariant formulation is inferior to its corresponding invariant covering theory. The final irony is that if there were justice based on chronological priorities Maxwell's theory would have been discarded even before Hertz's was discarded— and on identical grounds, namely, that of hair-trigger observational "disproof." For we mentioned that Mascart[2.8] (in 1872) and others had done experiments looking for Maxwell's predicted first-order fringe shifts and not finding them. So, Maxwell's theory was observationally "discredited" long before Hertz's, fully as firmly as Hertz's, and on identical grounds, *viz.*, a bad "ether" interpretation ruined it as physics. Still, as a fact of history, Maxwell's inferior formalism was retained and Hertz's superior formalism was discarded and utterly forgotten. (If extra irony is needed on top of irony, Hertz himself was quoted as doing Maxwell the favor of saying "Maxwell's theory is Maxwell's equations"—meaning, forget Maxwell's ether interpretation—yet no physicist ever did Hertz the favor of saying "Hertz's theory is Hertz's equations"—meaning, forget Hertz's ether interpretation!)

That's the way history of science works, and you won't find it in history of science books. Most of the latter are written to specs of political correctness of their day. They themselves are part of the history of science, and not the most admirable part. Still, it would be difficult to get along without them. For instance, the work of A. I. Miller,[2.2] an historian of science who yields place to nobody in his abject fealty to Einstein, is invaluable for giving a rare glimpse of the electromagnetic science that preceded Einstein. Thus, Miller's Eq. (1.8), attributed to Hertz, is identical to our Eq. (1.7), which we asserted to be demanded by Faraday's empirical evidence. So, although Miller does not make the point, Hertz's grasp of Faraday's physics—of the empirical basis of the budding electromagnetic science—was superior to that of the Maxwellians.

However, Miller's ideological bias in favor of his hero is so strong that he cannot resist making serious expository mistakes. He says of Hertz that "his axiomatic assertion of the form invariance of the electromagnetic field equations [or "covariance" as Hermann Minkowski (1908a) described this mathematical property] led Hertz to predict new effects whose empirical confirmation could in turn serve to confirm his axiom of covariance." This short quotation is riddled with misinformation. Hertz did not make an "axiomatic assertion." He simply and explicitly alleged the *invariance* of his equations (an allegation factually correct at first order, as we have seen). Unfortunately he left it at that, not bothering to prove his allegation—probably because he considered the reader intelligent enough to provide his own proof. Hertz rightly said nothing about covariance, because it had not been invented, was irrelevant, and was inapplicable to his equations. Those equations were, exactly as he said, *invariant*, not and never *covariant*. Concerning the vast physical difference between these two kinds of form preservation we shall have more to say below. The mathematical difference is equally vast, as should be evident to any reader who has had the mathematics of *covariance* drilled into him in the course of a modern higher education.

2.3 Invariance *vs.* covariance: The physics of it

It is worth pausing to say a few words about the neglected physical aspect of the distinction between *covariance* and *invariance*. It might be thought that there is no physics involved and that the distinction is purely mathematical—and not much of a distinction at that, since authorities as distinguished as Dirac slur over the difference by alluding to "Lorentz invariance." This writer, in fact, has never seen a clear, precise, mathematically kosher statement of the difference between invariance and covariance. So, I shall slur a bit myself, simply saying that under an *invariant transformation* a mathematical relationship (equation) preserves each symbol in place, from its untransformed to its transformed interpretation, without altering the formal relationships among symbols and without redefining individual symbol meanings. A *covariant transformation*, in contrast, allows similar form preservation of the relationships among symbols, but does so by introducing separate explicit relations of dependence (linear, in the case of Lorentz covariance) among transformed and untransformed symbols, amounting to *symbol redefinitions*. Symbol redefinition may be said to be the hallmark of "covariance."

Examples: The Maxwell field equations (1.1) are *covariant* under a Lorentz transformation, Eq. (1.4). (The explicit relations of linear dependence among field components are given in any of the texts, such as Jackson[1.1] or Panofsky and Phillips.[1.5] In effect there is a "scrambling" of electric and magnetic field components, such that electric field components in one frame are redefined as a linear combination of both electric and magnetic components in the other, *etc.*) The Hertz field equations (2.4) are *invariant* under the Galilean transformation (1.2), as we showed. Their symbols transform in place without redefinitions. $\vec{E}, \vec{B}$ in one frame mean exactly the same thing in any other, as per Eq. (2.6). It is my bias to characterize covariance as a clever contrivance, invariance as mathematically straightforward. (The symbol-redefinition approach succeeds through achieving *self*-consistency—and that in itself seems to me potentially dubious, in the sense that it would be better not to have to rely on it as physics. Mathematicians, in contrast, view self-consistency as self-justifying. Mary's little lamb, physics, is supposed to follow. Like the French king who said *"L'état, c'est moi,"* or the American Admiral who calls himself "Sixth Fleet," the mathematician declares, "Physics, it is my sym-

bols ... and my symbols are what I define them to be.") Both invariance and covariance exploit the theme of "form preservation," and on this basis both can lay formal claim to expressing a
physical relativity principle.

But physically the two describe entirely different sorts of
"fields," and no confusion should be allowed to exist between the
Maxwellian and Hertzian types. The easiest way to recognize the
difference is to contrast the ways in which the two are measured.
In Einstein-Maxwell theory each of an infinite ensemble of inertial observers has his personal point-like measuring instrument (a
multi-purpose type capable of measuring field components,
charge and current densities, *etc.*, in the manner indicated above)
at rest at his own field point, which is fixed in his system. At the
instant these field points coincide (the points being located on
lines in space parallel to the relative motions, all intersecting to
allow such coincidence), each of the measuring instruments ("detectors" brought into coincidence) is read by its respective observer and the various sets of field-component numbers so re-
corded are found on comparison to satisfy the relations of linear
dependence mentioned above and specified by the Lorentz transformation (with group parameter v descriptive of the relative motion of any pair of these inertial systems). Thus *operational procedures* or "measurements" are replicated in the various inertial systems, with the unspoken assumption (basic to all classical physics) of physical *replicability of experiments*.

Notice that this entails a collision of the various detectors, if
those are constructed of matter. To avoid this, one could picture
"near misses" of the detectors. But that is a compromise with operationalism. It would be more instructive to recognize that
Maxwell's conception of "field" as *Ding an sich* comes into play
here, so that Maxwellianism, *sang pur*, does not base its conceptu-
alizations upon measurement at all (or upon the pesky material
mechanisms of measurement), but simply upon abstract numbers
floating in space (attached to an independently existing "field"
reality) that are revealed by the equations without need for
measurement, even in thought. Hence in inertial transformations
the "collision" (harmless and non-destructive) is of mathematical
field points only, not of actual material (or notional) field detectors ... so there is no need to arrange near misses. This of course
does not fit with Einstein's working version of operationalism,
whereby everything hangs on *Gedanken* measurements, so it is

anomalous that Einstein took unquestioning satisfaction in exactly preserving Maxwell's formalism, root and branch. One
wonders if he understood the full philosophical implications of
what he was preserving.

So much for the Maxwell-Einstein form of electromagnetic
field theory. Let us turn next to the Hertzian version. Instead of a
raft, passel, or pride of inertial observers, each equipped with his
own "personal" point detector permanently fixed at his field
point, there is only a single point detector present—and this can
be unequivocally, without apologies or evasions, a genuine composition of matter. It is public property. Differently-moving inertial observers, of which there may be any number present, are all
provided, not with any tangible apparatus, but with a (unique)
numerical parameter value $\vec{v}_d$ descriptive of the velocity of this
single "public" detector relative to their own inertial system.
There is never any collision of real or notional detectors, because
there is only one detector present in the whole universe of discourse … and each observer is at rest with respect to his own
field point, as in Maxwell's theory. Each field point is in motion
aligned to pass through the location of the point detector. At the
moment when all field points coincide with the point detector's
location, all observers consult the read-out of *that single (public)
instrument*. Since they all read the same numbers from the same
instrument at the same instant, it follows that numerical invariance and form invariance are trivially satisfied. There is only one
physically unique *point event* of detection. That is what all ob-
servers describe. Eq. (2.6), $\vec{E}' = \vec{E}$, $\vec{B}' = \vec{B}$, just expresses this necessary identity of numbers read from the same instrument at the
same time by different observers. So, genuine invariance is a *Kin-
dergarten* matter—as distinguished from covariance, which is for
fast-track college sophomores and other candidates for initiation
into the Orphic and Eleusinian mysteries of higher mathematics,
who—trudging valiantly toward general relativity theory—bear
'mid snow and ice their banner with a strange device, "*Excelsior!*"

2.4 Invariance or covariance: Which is physics?

Is it a Mexican standoff, then, between invariant and covariant
formulations of electromagnetism? Is there no way to decide
which is physics? Will either describe all observable aspects of nature? At the level of classical description, in what may be called

the strong-field approximation (many photons present to represent the "field") this is undeniably the case. Maxwell's theory and its Hertzian covering theory do entirely equivalent jobs. Until about 1925 physicists had no reason to think in other terms. But then came quantum mechanics to overturn all ideas of the replicability of experiments on the micro scale. Recall that the Maxwell-Einstein model requires each member of an ensemble of inertial observers to be equipped with his own macro field-detection instrument, at rest at his field point, and all to replicate experimental procedures. This, as we noted, causes a conceptual problem of "collisions," when the covariant measurement is made at the instant of coincidence of the field points. To dodge that, we babbled of "near misses," but in the weak-field case all such gestures of accommodation fail.

For, consider the single-photon limit: In the simplest case we have two Maxwell-Einstein inertial observers with their macro-instruments coming together to make a simultaneous "measurement" of this photon. Do they both succeed? Of course not. The photon, by the basic nature of any "quantum" process, can be absorbed in at most one macro-detector. On a random basis, one of the two detectors present, let us say, "wins" and makes a successful measurement of the photon's "field." The other then necessarily "loses" and measures *zero*—a big goose-egg! Experimental replicability thus fails catastrophically. Turn and twist as he may, there is no way any self-respecting mathematician, not to mention physicist, can represent the one observer's "zero" as related by the *Lorentz transformation* to the other observer's finite measured values of the field components. Hence covariance clearly and manifestly fails irreparably in the weak-field limit. The idea of *universal covariance* is a joke, a propagandist's mantra, an unchecked flowering of *New York Times* science. It does not work even within all electromagnetism, much less within all nature.

Does invariance succeed in this limit? Of course it does, trivially so—because there is only a *single* Hertzian macro-detector present. That either detects the photon or it doesn't, on a random basis. The fraction of trials on which it succeeds is related to the Born (probabilistic) interpretation of the photon's wave function. All very neat—no problems, because no "interference" or "competition" among macro detection devices for the weak-field quantum. The upshot, then, is that Hertzian invariance works in both strong- and weak-field cases, whereas Maxwell-Einstein covariance works only for strong (or "classical") fields.

If physics wants to progress smoothly, with integrity and consistency, into the quantum descriptive domain, it must drop covariance at the threshold and switch to invariance. (Having done that, it might as well drop covariance, period ... and all the elaborate cloud-castle of twaddle with which it is laden.) This is such a simple and irrefutable argument that I have found it to beget an equally simple and irrefutable scientific response: universal ignoration. The Big Yawn.

Such a response may be bolstered by the modern trend of philosophers and philosophically inclined physicists to view operationalism as *discredited*. The urge to see things as either credited or discredited for good and all is one of the subliminal influences toward simplism that prevent much net progress of scientists *en masse* toward adulthood. If physics is to be about what is "measurable," it needs to be able to give at least a conceptual or *Gedanken* account of how the necessary measurements are made. That is what I have insisted upon here. I call this "instrumentalism" or "operationalism." Natural philosophers and scholars of science are notable for sicklying o'er with the pale cast of thought whatever they view as fundamental in the description of nature. I decline to chop logic with them about the sicklying job they did in allegedly "discrediting" operationalism. But I do want to know: Where have they been all these years since that hatchet job? Have they mislaid their sicklying gear? Why no sicklying o'er of the problem of electromagnetic measurement? How has field theory escaped so long any mobilization of scholarly critical apparatus capable of probing just how (by what instruments in what states of motion) the "field" is measured? In other words, how has covariance, in its imperial nakedness, so long escaped being laughed at? And for how much longer? Surely modern scholarship is itself some kind of joke—another weapon in the arsenal of political correctness. We seem to have built the kind of world in which only cynics know how or when to laugh.

2.5 Hertzian wave equation

The first-order invariant wave equation is derived in the customary way: Taking the curl of Eq. (2.4a), we have in the source-free case $\left(\vec{j}_m = 0 \right)$

$$\vec{\nabla}\times\left(\vec{\nabla}\times\vec{B}\right)=\frac{1}{c}\vec{\nabla}\times\frac{d\vec{E}}{dt}=\frac{1}{c}\frac{d}{dt}\vec{\nabla}\times\vec{E}=-\frac{1}{c^2}\frac{d^2\vec{B}}{dt^2},$$

this last equality following from taking the total time derivative of Eq. (2.4b). Applying a standard vector identity,[2.3] we have

$$\vec{\nabla}\times\left(\vec{\nabla}\times\vec{B}\right)=\vec{\nabla}\left(\vec{\nabla}\cdot\vec{B}\right)-\nabla^2\vec{B}=-\frac{1}{c^2}\frac{d^2\vec{B}}{dt^2},$$

where $\nabla^2=\vec{\nabla}\cdot\vec{\nabla}$; and, since $\vec{\nabla}\cdot\vec{B}=0$ from (2.4c), the Hertzian wave equation—invariant at first order under Galilean transformations—is

$$\nabla^2\vec{B}-\frac{1}{c^2}\frac{d^2\vec{B}}{dt^2}=0. \qquad (2.17)$$

Similarly, taking the curl of Eq. (2.4b) and the total time derivative of (2.4a) for $\vec{j}_m=0$, we derive

$$\nabla^2\vec{E}-\frac{1}{c^2}\frac{d^2\vec{E}}{dt^2}=0. \qquad (2.18)$$

Taken together, (2.17) and (2.18) describe electromagnetic wave propagation in free space (vacuum). These wave equations are invariant at first order under the Galilean transformation (because derived from field equations already shown to be invariant). They possess this property by virtue of our having substituted the first-order invariant operator d/dt for the traditional Maxwellian non-invariant $\partial/\partial t$. From this substitution follow interesting physical consequences, several of which we shall presently explore.

Although the quantity c appearing in (2.17) and (2.18) is invariant, in agreement with (2.13), this does not imply that the physical speed of light is invariant. To explore this, we must solve the Hertzian wave equation. Applying standard field theoretical analytic techniques, let us look for a solution of (2.18) of d'Alembertian type. (For simplicity we may confine attention to the electric field $\vec{E}$, since the results for $\vec{B}$ are formally identical.) The solution we seek is of the form $\vec{E}=\vec{E}(p)$, where

$$p=\vec{k}\cdot\vec{r}-\omega t=xk_x+yk_y+zk_z-\omega t. \qquad (2.19)$$

We find that

$$\nabla^2\vec{E}(p)=\left(\frac{\partial^2}{\partial x^2}+\frac{\partial^2}{\partial y^2}+\frac{\partial^2}{\partial z^2}\right)\vec{E}(p)$$

$$=\left(k_x^2+k_y^2+k_z^2\right)\vec{E}''(p)=k^2\vec{E}''(p), \qquad (2.20)$$

where double prime denotes two differentiations with respect to p. Similarly, using Eq. (1.8), we have

$$\frac{d^2}{dt^2}\vec{E}(p) = \left(\frac{\partial}{\partial t} + \vec{v}_d \cdot \vec{\nabla}\right)^2 \vec{E}(p)$$

$$= \left(\frac{\partial^2}{\partial t^2} + 2\frac{\partial}{\partial t}(\vec{v}_d \cdot \vec{\nabla}) + (\vec{v}_d \cdot \vec{\nabla})^2\right)\vec{E}(p) \qquad (2.21)$$

$$= \left(\omega^2 - 2\omega(\vec{v}_d \cdot \vec{k}) + (\vec{v}_d \cdot \vec{k})^2\right)\vec{E}'' = \left(\omega - (\vec{v}_d \cdot \vec{k})\right)^2 \vec{E}''.$$

Eqs. (2.18), (2.20), (2.21) imply

$$\left[-k^2 + \frac{1}{c^2}\left(\omega - (\vec{v}_d \cdot \vec{k})\right)^2\right]\vec{E}'' = 0. \qquad (2.22)$$

From the vanishing of the coefficient of $\vec{E}''$ it follows that

$$ck = |\omega - \vec{v}_d \cdot \vec{k}| \quad \text{or} \quad \frac{\omega}{k} = \pm c + \frac{\vec{k}}{k} \cdot \vec{v}_d, \qquad (2.23)$$

where $k = \sqrt{\vec{k} \cdot \vec{k}}$. It is useful to define a wave phase propagation speed ("phase velocity") u relative to the observer's inertial system, as

$$u = \frac{\omega}{k} = \pm c + \frac{\vec{k}}{k} \cdot \vec{v}_d. \qquad (2.24)$$

The corresponding result in Maxwell's theory is $u = \pm c$. Consequently we see that the universal constancy of light speed (Einstein' second postulate) is not satisfied in Hertzian theory—so a new kinematics will have to be devised to describe high-speed motions. That we leave for later chapters.

Evidently, the most general d'Alembertian solution of Eq. (2.18) can be written, from (2.19) and (2.24), as

$$\vec{E} = \vec{E}_1\left[\vec{k} \cdot \vec{r} + (ck - \vec{k} \cdot \vec{v}_d)t\right] + \vec{E}_2\left[\vec{k} \cdot \vec{r} - (ck + \vec{k} \cdot \vec{v}_d)t\right], \qquad (2.25)$$

where $\vec{E}_1, \vec{E}_2$ are arbitrary vector functions. Invariance of the field solutions, Eq. (2.6), implies that $\vec{E}(p) = \vec{E}'(p')$. This in turn implies a condition of phase stability, $p = p'$, or, from (2.19),

$$\vec{k} \cdot \vec{r} - \omega t = \vec{k}' \cdot \vec{r}' - \omega' t', \qquad (2.26)$$

which describes a constant value of the phase of the propagating wave field as measured in the primed and unprimed inertial systems. Suppose we limit attention to the case of uniform motion of the field detector. Then we can specialize by considering the detector at rest in the primed inertial system, so that $\vec{v}_d = \vec{v}$, where $\vec{v}$ is the (constant) velocity of the primed system with respect to the unprimed one. In this case, eliminating $\vec{r}', t'$ from (2.26) by means of the Galilean transformation, Eq. (1.2), we find after rearrangement that

$$\left(\vec{k} - \vec{k}'\right) \cdot \vec{r} = t\left(\omega - \omega' - \vec{k}' \cdot \vec{v}_d\right). \tag{2.27}$$

Since $\vec{r}, t$ are arbitrary and independently variable, their coefficients must vanish. Consequently,

$$\vec{k}' = \vec{k} \tag{2.28}$$

and

$$\omega' = \omega - \vec{k} \cdot \vec{v}_d. \tag{2.29}$$

The first of these results describes first-order aberration, the second the first-order Doppler frequency effect of detector motion. Since stellar (Bradley) aberration is ordinarily described as an apparent small turning of the $\vec{k}$-vector of starlight propagation, we see that no such apparent turning is predicted by Eq. (2.28), which asserts first-order invariance of the $\vec{k}$-vector. Starlight from the pole of the ecliptic [for which $\vec{k} \cdot \vec{v}_d = 0$, hence $|u| = c$ from (2.24)], cannot depart from constancy in either speed or direction. A direct way to describe stellar aberration will therefore involve a higher-order refinement of the Hertzian theory, which will be our topic in Chapter 3. Another way will be discussed in Chapter 7. This $\vec{k}$-constancy consideration alone is sufficient to show that unadorned Hertzian theory, as most simply interpreted, is inadequate to the physics and needs higher-order refinement. We have worked it out here in detail to show how a self-consistent first-order invariant treatment can be developed formally—as well as to provide a model for the higher-order calculation.

The reader should not be disturbed that first-order electromagnetic theory is unable to handle stellar aberration. [Maxwell's theory does no better, since it shares Eq. (2.28).] SRT claims to handle it (about that claim we shall have more to say in Chapter 4), but it is a second-order theory. To give invariance an equal chance to do the job, we owe it a second-order development. The phenomenon *appears* to be a first-order effect because of proportionality of the observed aberration angle to a velocity. But stellar aberration is in fact a much subtler phenomenon than it seems or was thought to be classically. We shall argue in Chapter 4 that what it involves is not a directional turning of the $\vec{k}$-vector but a second-order effect of detector motion on light speed. Because of the one-way light propagation involved, stellar aberration also represents a much more important and revealing aspect of the physics of light than is generally recognized.

2.6 Potier's principle

Before leaving the first-order development of invariant field theory, it is necessary to revert to some nineteenth-century lore that has a direct bearing on why the bizarre and counter-intuitive "convection of light by the absorber," apparently predicted by our wave equation solution (2.24), is not in general observable. As we study Eq. (2.24), we see that it suggests a very counter-intuitive thing—that motion of the field detector or photon absorber with respect to the observer *changes the speed of light*, by pulling the photons along with it—as it were, *convecting* them. If the direction of light propagation, specified by $\vec{k}/k$, is parallel to detector motion $\vec{v}_d$, then light speed ("phase velocity") according to (2.24) is $c + v_d$, and if anti-parallel it is $c - v_d$. [This tacitly assumes the choice of plus sign for $\pm c$ in (2.24).] It might be thought that such a grossly acausal notion (after all, how does the photon know about its future absorber's motion before it has even reached that absorber?) is directly counter to experience, so the whole Hertzian fabrication can be dismissed. However, phase velocities (like photon trajectories and mental "pictures" of light propagation or other quantum processes generally) are without directly observable counterpart in nature. Moreover—and this is the point of immediate consequence—there is a theoretical result of nineteenth-century physics, known as Potier's principle, which makes it understandable that the simplest forms of first-order observations, such as ordinary laboratory fringe shifts, can reveal no observable effect on light speed of any additive term of type $\vec{k} \cdot \vec{v}$ entering the expression for phase velocity.

Actually, to be true to history, we should say that Potier's principle was aimed at ether theories. It showed that fringes would not be shifted by any additive phase velocity term of the form $\vec{k} \cdot \vec{V}$, if $\vec{V}$ were interpreted as ether velocity. But the mathematics of Potier's proof applies to any "$\vec{V}$," regardless of how interpreted physically. So, it applies here with $\vec{V}$ interpreted as detector velocity $\vec{v}_d$. The principle is discussed in several modern references, for instance.[2.9-2.11] It will be derived here for the convenience of readers to whom such references are not available. Nineteenth century geometrical optics is assumed, but presumably a modernized wave formulation would not alter the gist of the treatment or result.

Potier's principle:[2.12] Let two fixed points in an inertial system, P₁ and P₂, of emission and absorption, respectively, be joined by a light path consisting of n_0 connected straight-line segments, the ith one of which may be denoted vectorially as $\vec{\ell}_i = \ell_i \vec{\varepsilon}_i$, where $\vec{\varepsilon}_i = \vec{k}_i / k_i$ is a unit vector of light propagation on the ith segment, $i = 1, 2, \ldots, n_0$. Let the index of refraction in the vicinity of the ith path segment be given by a law of the form $n_i = 1 + a\vec{\varepsilon}_i \cdot \vec{V}\left(1 + O(V/c)\right)$, where a is a scalar constant and $\vec{V}$ is a constant vector. [Here $O(V/c)$ denotes any quantity of the order of V/c in magnitude, considered small.] Then

(a) The total light transit time from P₁ to P₂ is increased (compared to the case $V = 0$) by an amount $(a/c)\vec{L} \cdot \vec{V}\left(1 + O(V/c)\right)$, where $\vec{L} = \sum_{i=1}^{n_0} \vec{\ell}_i$ is the vector (straight-line) distance from P₁ to P₂.

(b) The spatial path taken by physical light in passing from P₁ to P₂ is unaffected by arbitrary changes in $\vec{V}$ (for $V \ll c$).

Proof: The time required for light to transit the ith path segment is

$$t_i = \frac{\ell_i}{u_i} = \frac{\ell_i}{(c/n_i)} = \frac{\ell_i}{c}n_i = \frac{\ell_i}{c}\left[1 + a\vec{\varepsilon}_i \cdot \vec{V}\left(1 + O(V/c)\right)\right]$$

$$= \frac{\ell_i}{c} + a\ell_i\vec{\varepsilon}_i \cdot \frac{\vec{V}}{c}\left(1 + O(V/c)\right) = \frac{\ell_i}{c} + \frac{a}{c}\vec{\ell}_i \cdot \vec{V}\left(1 + O(V/c)\right).$$

Hence the total transit time from P₁ to P₂ is

$$T(\vec{V}) = \sum_{i=1}^{n_0} t_i = \frac{1}{c}\sum_{i=1}^{n_0} \ell_i + \frac{a}{c}\left(\sum_{i=1}^{n_0} \vec{\ell}_i\right) \cdot \vec{V}\left(1 + O(V/c)\right)$$

$$= T(0) + \frac{a}{c}\vec{L} \cdot \vec{V}\left(1 + O(V/c)\right),$$

(2.30)

where $T(0)$ is the transit time for $\vec{V} = 0$. This proves part (a) of the principle. To establish part (b) we invoke Fermat's principle, which states that for arbitrary path variations in close proximity to the physical light path P, connecting fixed end points P₁ and P₂, the time of light transit is an *extremum* (least) for the actual physical path P. We have just shown that total light transit time (the only observable) is determined by the constant vector $\vec{L}$ between fixed endpoints P₁ and P₂, and is independent of the index i, hence of the particular values of the $\vec{\ell}_i$; that is, independent of the detailed geometry of the path between those endpoints. Hence for an arbitrary constant value of $\vec{V}$

$$T(\vec{V}) = T(0) + (Constant\ for\ all\ varied\ paths). \qquad (2.31)$$

Since $T(0)$ does depend on path and is a minimum on path P, it follows that $T(\vec{V})$ is also a minimum on P for arbitrary $\vec{V}$. Therefore the physical path of light—that which takes the least time—is the same for $\vec{V} \neq 0$ as for $\vec{V} = 0$, q.e.d.

It will be understood that the statement here of Potier's principle in terms of an "index of refraction" is arbitrary and does not follow history. (Potier's original formulation treated "ether" motion in the context of Fresnel's theory.) The above form of the principle can be connected to the present work by reference to Eq. (2.24). Choosing the plus sign for c in that equation and defining an index of refraction of free space by

$$n = \frac{c}{u} = \frac{c}{c + \dfrac{\vec{k}}{k} \cdot \vec{v}_d} = \left(1 + \frac{\vec{k}}{kc} \cdot \vec{v}_d \right)^{-1} = 1 - \frac{1}{c}\frac{\vec{k}}{k} \cdot \vec{v}_d + O\left((v_d/c)^2 \right), \ (2.32)$$

we see that this satisfies the conditions of the theorem, given the identifications $a = -(1/c)$, $\vec{\varepsilon}_i = \vec{k}/k$ for all i, and $\vec{V} = \vec{v}_d$. So, just as ether wind is unobservable in the spatial domain at first order, the same is true of our apparent phenomenon "convection of light by the absorber." (Formally, our identification $\vec{V} = \vec{v}_d$ implies that we are thinking of the "ether" as convected by the absorber. That is, the light detector is always at rest in the luminiferous medium, or carries the medium with it.) The index of refraction of free space is indistinguishable from unity (the value for $\vec{v}_d = 0$) by any experiments in the spatial domain (interferometry, diffraction, etc., related to "path"). This conclusion is as valid as, and is founded upon, Fermat's principle. Historically, Potier's principle established the futility of all experimental attempts to challenge the relativity principle at first order. It was doubtless a major motivator of Michelson-Morley's investigation at second order.

To interject a personal note, I may remark that I wasted several years of experimental effort looking for first-order effects of "light convection by the absorber," using interferometry, diffraction gratings, etc., all with null results. These trials are summarized in Chapter 7 of *Heretical Verities*.[2.11] At the time I knew of Potier's principle, but failed to make the simple connection from ether convection to absorber-motion convection. It is sobering to find years of laboratory labor obviated by a few minutes of thought. The same might prove true of theoretical labor, if this lesson were to be applied more generally.

Does this mean there is no hope of observing any distinction between invariant Hertzian and covariant Maxwellian theories of electromagnetism? No, it merely means that effects of the $\vec{k}\cdot\vec{v}_d$ phase term will not be directly observable as a basis for the distinction. Furthermore, part (a) of Potier's principle leaves open the possibility of experiments in the time domain. In analyzing stellar aberration in Chapter 4, we shall find that the Hertzian second-order phase term has a first-order effect on the observable phenomenon. Consequently, stellar aberration will play a stellar role in deciding empirically between the Hertzian and Maxwell-Einstein views of field theory.

2.7 Sagnac effect and ring laser

In the Sagnac experiment the detector moves circularly in an inertial system, somewhat as in stellar aberration, so one might hope from the foregoing that such an experiment would cast light on the issue of Hertzian *vs.* Maxwellian electromagnetism. Unfortunately, Potier's principle spoils this hope, as we shall see. In the Sagnac apparatus monochromatic light circulates around a planar area A of any shape in opposite directions on a platform, establishing interference fringes. The platform rotates, along with A, at angular velocity ω about an axis normal to A located anywhere (inside or outside A). As a result of the rotation either of two results is observed in practice,

(a) zero fringe shift relative to the case of no rotation,
(b) a first-order fringe shift of magnitude consistent with a circulation time discrepancy between the two beams of $\Delta t = 4A\omega/c^2$, this latter result being evidence of the "Sagnac effect."

What makes the difference between these two cases? Detailed experimental conditions, seldom discussed. If, as in the original experiments of Sagnac, there is a good deal of vibration, or if there is deliberate "jitter" introduced into the light propagation process, result (b) is observed. If the apparatus is comparatively steady and no dither is employed, result (a) is apt to be obtained.

A jargon has been developed to describe the physics of the latter case; *viz.*, it is called "frequency pulling," resulting in "mode locking." Frequency pulling refers to an empirically observed tendency of the light frequency to shift spontaneously to the nearest *eigen*-frequency such that an integral number of wavelengths fits the cavity length (or light path optical length). The cavity then behaves as if "sticky" and carries the standing wave pattern along with it without a Sagnac shift, as if the apparatus were not rotating. To cause this phenomenon of mode locking, some interaction between the two counter-propagating modes must be present, known as mode coupling, and indications are that the coupling agency is generally the backscattering of light from imperfectly reflecting mirrors. Although mirrors of reflectance as high as 99.9999% have been used, there is always some mode locking. Kelly[2.13] (page 52 of his book) comments as follows:

> There was a considerable element of luck in the original Sagnac experiment. It was later found that the light signals can lock onto the circuit and mirrors unless there is considerable vibration; such vibration was present in the Sagnac experiment. In later designs a dither is introduced to ensure that locking does not occur. History would, no doubt, have taken a different turn had the Sagnac test given a zero result, which would have been the case had the equipment been rock steady ...

The dithering introduced to combat mode locking may involve a small superposed twisting oscillation about the rotation axis, or some other ingenious form of perturbation. Physicists view the phenomenon as not of fundamental physical interest, but merely as an apparatus effect to be overcome. For purposes of present discussion, I am willing to go along with this and to accept that the "real physics" is result (b), above. Let us turn to theory, then, and see how it explains that outcome.

First, what does Maxwell-Einstein theory predict about the Sagnac experiment? Identifying the laboratory as close enough to an inertial system, and declaring light always to have speed c in inertial systems, they note that light circulating at that speed relative to the laboratory in the direction of platform rotation will have to go an extra increment of laboratory *travel distance* Δs in order to "catch up" in completing the light circuit, whereas light circulating counter to the sense of rotation will travel a shortened distance Δs in returning to its starting point. [It is easily shown for a circular path rotating about its center that Δs bears to path length the ratio $(v/c)^2$. However, $\Delta t = \Delta s/v$, the time (or phase) discrepancy, is of first order in v.] When the beams reunite and interfere to form fringes there will be a travel time discrepancy Δt between them, and the first-order fringe shift of outcome (b) is expected to be observed. This is the prediction of SRT—and of course of classical first-order kinematic theory as well. All that's really involved is a small "travel distance" discrepancy in the lab for the counter-circulating beams. The physics behind this seems to be that optical phenomena, like mechanical ones, need to be analyzed in inertial systems to yield simplest descriptions. This necessity appears to depend only on the Newtonian theoretical framework.

The fact that there is a first-order travel time discrepancy Δt observed by the rotating platform rider R as well as by the lab observer—both observing the same factual phase discrepancy or fringe shift—suggests that R must attribute different speeds to light going in the two directions, since the two beams, starting simultaneously and traveling equal total first-order distances in the rotating system, come back to the same place in that system at different times ... but many Einsteinians cannot bear this implication, because it asserts a first-order non-constancy of light speed in a non-inertial system. Other Einsteinians take it in stride, accepting (Einstein, *The Meaning of Relativity*, 1922) that "vacuum light is propagated with the velocity c, at least with respect to a definite inertial system."

Now let's look at the same picture from the standpoint of Hertzian theory. As before, our observer is at rest in the inertial laboratory, but we note that the light path contains a number of mirrors, optical fibers, light pipes, or other tangible objects in rotary motion that may be viewed as "detectors" or absorber-re-emitters of the light. A typical one of these, moving at velocity $\vec{v}_d$ in the laboratory, will formally convect the light traveling with it, in accordance with Eq. (2.24)—imparting relative to the laboratory a phase speed $u = \pm c + (\vec{k}/k) \cdot \vec{v}_d$. However, Potier's principle applies (given that we conceive of this as a fixed endpoint problem, owing to the fact that the *inertial observer's field points*, source and sink, are indeed fixed in the lab, regardless of detector motion) and we are forced to conclude that only the $u = \pm c$ part of this is observable, as in Maxwell's theory. Consequently, with effective speed c in the laboratory, we are back to the classical travel distance Δs effect, previously discussed, as the only first-order observable influence on the fringes. Such reasoning predicts outcome (b), the Sagnac effect. The latter, then, is disqualified as a means of deciding between Maxwell and Hertz. If the rotating platform rider R attempts to use Hertzian theory he is also no better off than the Maxwellian observer, since for both the detector is at relative rest $(\bar{v}_d = 0)$ and the two theories become equivalent. Thus both teach the lesson mentioned above, that optical problems, like mechanical ones, require analysis with respect to an inertial system for simplest description.

The ring laser (often used as a frictionless gyroscope) differs in significant details from the Sagnac interferometer. In most cases no dithering is employed, and the observable quantity is not a fringe shift but the frequency of passage past a fixed observation point of moving "beats" of a standing-wave pattern set up in a fixed optical fiber, light pipe, or other light-conducting circuit, the whole together rotating as a rigid unit at angular frequency ω in an inertial system. The formula describing this phenomenon is $df = 4A\omega/\lambda P$, where A is the area of the optical circuit, P the optical (not physical) path length of its perimeter, λ the light wavelength, and df the frequency at which nodes of the *moving* "standing wave" beat pattern pass the observation station at any fixed point along the circuit. Because of the appearance of a geometry-dependent factor A/P in the formula, the observed df depends on the shape of the circuit ... but for all shapes and sizes the observed df is proportional to ω. This method has been used to improve (vastly) on the accuracy of the Michelson-Gale[2.15] measurement of the spin angular velocity of the earth (which employed a large-scale Sagnac interferometer).

In the active form of the ring laser the cavity, generally a low-pressure gas-containing Pyrex tube, square or triangular in planar configuration, with high-reflectance mirrors at the corners, is caused to "lase" by electromagnetic excitation. Beats of the resulting standing wave pattern move at the frequency df past a detection station fixed with respect to the tube. A frequency-pulling phenomenon in such geometry is observed only for small enough df values, below a certain "lock-in threshold." Generally, no attempt is made to combat this; but the threshold is made very low, even into the micro-Hertz range, by using nearly perfect mirrors to reduce backscatter. Although the experimental details differ between the ring laser and the classic Sagnac interferometer, the gist of the analysis is the same. Hertzian theory, as above, is predictively equivalent to Maxwellian theory.

Different versions of the Sagnac experiment have been carried out in which the light source rotates with the rotating platform or rests in the laboratory. Similar fringe shifts are observed in the two cases. This confirms that source motion is irrelevant, just as the relativists maintain. Further variants of the Sagnac idea include fiber-optic gyroscopes, fiber-optic conveyors, *etc.* A wealth of reference material is accessible on the Internet. The use of fiber optics allows easy exploitation of Sagnac's recognition of the arbitrariness of optical circuit shape. Thus Wang *et al.*[2.14] have distorted the circuit shape into linearity on most of the circuit, so that the fiber motion can be made mainly inertial, parallel and anti-parallel to light propagation direction. These workers have demonstrated interference fringe shifts of counter-propagating light beams in such fiber optic circuits, major portions of which move uniformly and rectilinearly in the laboratory.

Experiments of this kind are in a way analogous to the Faraday experiment in which a portion of an electrical circuit was moved in a magnetic field. As in that case, they defy conventional SRT analysis because of lack of an inertial frame in which the circuit as a whole can be considered to be at rest. Wang's reported empirical finding was an optical fringe shift $\Delta\phi = 4\pi\vec{v}\cdot d\vec{\ell}/c\lambda$, which is a linear expression of the Sagnac formula in differential form. Wang was under the impression that his observations (of fringe shifts in optical circuits consisting mainly of inertial portions, with translatory motions instead of rotary) refuted Einstein's second postulate of light-speed constancy, inasmuch as a shift must correspond to equal travel distances of counter-circulating light beams in unequal times (implying light speed non-constancy in inertial systems). But this can doubtless be countered by the observation that fringe formation requires circuit completion, and the circuit as a whole is not at rest in any single inertial system. When the Lorentz transformation event calculus is applied to counter-moving inertial systems, it will doubtless turn out that the first-order situation is saved for SRT by what amounts to a disguised version of the classical "travel distance" argument already rehearsed. Still, although SRT may not be refuted, the physical motions in Wang's experiment are essentially inertial, and one becomes painfully aware of the possibility of other, less contrived, explanations of the observations. Indeed, as Stedman[2.16] remarks with a straight face, "it is now generally recognized that the prediction of [the Sagnac formula] is remarkably robust to the assumed theoretical framework .... "

2.8 A bit of GPS evidence

The Global Positioning System (GPS) provides further evidence related to the Sagnac effect and to other topics that will interest us. We shall have more to say about it in subsequent chapters. Dr. Ron R. Hatch, a GPS expert, sent me the following e-mail on 24 August, 2005:

> Using the GPS system and the clocks set to run at a common earth surface time ... The range to the GPS satellites is computed from the transit time from satellite to ground. The transit time is multiplied by the speed of light to get the range. But the speed of light has to be adjusted by the component of the receiver velocity (including the earth spin) away or toward the satellite. Thus, the speed used is $(c + v)$ and $(c-v)$. They call it a Sagnac effect even though the deviation from a straight line during the transit time is on the order of 10^{-11} meters. I have argued the circular path has nothing to do with it and in the latest GPS ICD they admit that any receiver motion in addition to the rotation must be used in the adjustment.

It is to be noted that the velocity "v" used in the GPS calculations just mentioned is referred to an inertial system S moving with the center of a non-rotating earth. In that system the first-order Hertzian phase velocity of light, Eq. (2.24), is $u = \pm c + \left(\vec{k}/k\right) \cdot \vec{v}_d \to c \pm v_d$, where $\vec{v}_d$ is the velocity of a detector at rest or in motion on or near the surface of the spinning earth. The field point is at rest in S, so Potier's principle (applicable alike to one-way and two-way paths with fixed endpoints) again rules out any observable distinction between Hertz's and Maxwell's theories. In both theories we are left with a distance effect Δs, as in the Sagnac effect ... although in this case because of linearity it might more aptly be termed a Wang effect. Again, the main lesson is that simplest descriptions require reference to a "sufficiently inertial" system.

In all these applications, what shall be meant by the assertion that a reference system is sufficiently inertial? The only possible answer is that circumstances alter cases. In the original Sagnac experiment, to good enough approximation, the inertial reference system was the laboratory. In the Michelson-Gale version the sufficiently inertial system was one moving with the center of a non-rotating earth. The same is true for the GPS. In the case of stellar aberration (*cf.* Chapter 4), since the tiny aberration phenomenon exhibits an annual period, a more rigorous criterion of "inertiality" applies, requiring reference to the barycenter of the solar system or equivalent. For very long-term observations, wherein the solar system deviates appreciably from linear motion, the barycenter of the galaxy might be needed.

2.9 Chapter summary

Hertz's version of first-order electromagnetic field theory, an invariant (under the Galilean transformation) covering theory of Maxwell's non-invariant version, has been sketched here. It failed historically because of a false "ether" interpretation, but when that is corrected very few observable departures from Maxwell are predicted. Appearances indicative of a gross Hertzian "convection of light by the absorber" prove deceptive, in that such putatively acausal effects are shown by Potier's principle (a consequence of Fermat's principle) to be unobservable at first order in somewhat the same sense that "ether wind" is unobservable. (That is, unobservable in the space domain, as by fringe shifts; the time domain being another matter, as yet unprobed in the laboratory.) The Sagnac effect, in its various guises—because of Potier's principle—provides no evidence to decide between the Maxwell and Hertz pictures of light propagation.

The issue of covariant (Maxwell) *vs.* invariant (Hertz) formulations of field theory, while not resolvable by ordinary laboratory observations, is settled on the side of theory through the intervention of quantum considerations: Invariant (single detector) theory handles both weak-field and strong-field cases, whereas covariant (multi-detector) theory works only in the strong-field limit. Covariance is thus discredited on its home ground, electromagnetism. Consequently, the historic favor of "discreditation" is returned, a century later, from the Hertz (German) to the Maxwell (English) political camp. We find, however, that the Hertzian wave equation, taken in its original context, fails to describe stellar aberration. For that, the inadequacy of a first-order formulation is responsible. A "neo-Hertzian" higher-order approximation, to be developed in the next chapter, will be found to do better.

References for Chapter 2

[2.1]   H. von Helmholtz, *Borchart's J. Math.* **78**, 273-324 (1874).

[2.2]   A. I. Miller, *Albert Einstein's Special Theory of Relativity Emergence (1905) and Early Interpretation (1905-1911)* (Addison-Wesley, Reading, MA, 1981).

[2.3]   J. A. Stratton, *Electromagnetic Theory* (McGraw-Hill, NY, 1941), App. II.

[2.4]   M. Abraham and R. Becker, *The Classical Theory of Electricity and Magnetism* (Blackie and Son, London, 1932), Vol. 1.

[2.5]   J. Dunning-Davies, *Progress in Phys.* **3**, 48-50 (2005).

[2.6]   H. R. Hertz, *Electric Waves*, translated by D. E. Jones (Dover, NY, 1962), Chap. 14.

[2.7]   A. A. Eichenwald, *Ann. Phys.* (Leipzig) **11**, 1 (1903); *ibid.*, 421.

[2.8]   E. E. N. Mascart, *Ann. École Norm.* **1**, 157-214 (1872); *ibid.*, **3**, 157-214 (1874).

[2.9]   R. G. Newburgh, *Isis* **65**, 379 (1974).

[2.10] R. G. Newburgh and O. Costa de Beauregard, *Am. J. Phys.* **43**, 528 (1975).

[2.11] T. E. Phipps, Jr., *Heretical Verities: Mathematical Themes in Physical Description* (Classic Non-fiction Library, Urbana, 1986).

[2.12] A. Potier, *J. de Physique* (Paris) **3**, 201 (1874).

[2.13] A. G. Kelly, *Challenging Modern Physics—Questioning Einstein's Relativity Theories* (BrownWalker, Boca Raton, 2005). Also A. G. Kelly, "Synchronization of Clock Stations and the Sagnac effect," in *Open Questions in Relativistic Physics*, F. Selleri, ed. (Apeiron, Montreal, 1998).

[2.14] R. Wang, Y. Zheng, and A. Yao, *Phys. Rev. Ltrs.* **93**, No. 14, 143901 (2004); also R. Wang, Y. Zheng, A. Yao, and D. Langley, *Phys. Ltrs. A*, **312**, 7-10 (2003) and R. Wang, *Galilean Electrodynamics* **16**, No. 2, 23-30 (2005).

[2.15] A. A. Michelson and H. G. Gale, *Astroph. J.* **61**, 137-145 (1925).

[2.16] G. E. Stedman, *Rep. Prog. Phys.* **60**, 615-688 (1997).

> ... the supreme goal of all theory is to make the irreducible basical elements as simple and as few as possible without having to surrender the adequate representation of a single datum of experience.
>
> —Albert Einstein, *On the Method of Theoretical Physics* (1933)

Chapter 3

Higher-order Electrodynamics ... (the neo-Hertzian Alternative)

3.1 The higher-order kinematic invariants

In the last chapter it seemed that an operationalist or instrumentalist philosophy could contribute to clear thinking by forcing us to examine the details of our measurement procedures—if only in *Gedanken* terms. We found that first-order electromagnetic theory could benefit immensely from applying such intellectual discipline. For instance, the reasoned contemplation of how measurements would have to be made ruled out spacetime symmetry and covariance (through consideration in the ultimate weak-field limit of the single quantum competed-for by a plurality of macro-instruments) in favor of true invariance. Let us see if a similar philosophy can guide us in the labyrinth of higher-order approximation.

The *leitmotif* of instrumentalism is the question, *how do we measure* whatever it is we recklessly conceptualize in our theorizing? Pursuit of that inquiry proves to be a relatively painless way of disciplining our idealizations. The young Einstein took the lead in asking that question about one of the most fundamental of physical descriptors, *time*. His answer hinged upon recognition of a crucial distinction between (a) inertial "frame time" *t*, measured by a spatially extended set of clocks at rest in that inertial frame and subject to a specified ("Einstein") synchronization convention, and (b) "proper time" τ of a particle in an arbitrary state of motion, measured by a single co-moving clock. In my opinion Einstein's most enduring and important contribution to relativistic physics was his definition relating the differentials of these two types of "time;" namely,

$$d\tau^2 \equiv dt^2 - dr^2/c^2 = invariant$$
$$dr^2 = dx^2 + dy^2 + dz^2. \tag{3.1}$$

It is understood that all differentials here refer to a pair of events occurring successively on the trajectory of a single (classical or macro) particle, as measured by instruments at rest in an inertial frame in the case of dr, dt, and by a clock at rest with respect to the particle in the case of $d\tau$. It is also understood that on this trajectory "other things" (such as gravity) are "equal."

This definition was actually a leap of inspiration, not a strict deduction from observation. It was, however, a most grand and far-reaching leap—and one for which I can only express admiration ... for it has been copiously confirmed in its implications by experiments of the sort for which I profess esteem. Indeed, I find it tempting to make a further leap and promote (3.1) to the status of *fact*. This is not because I think such a course logically defensible, nor acceptable as methodology, but simply because adopting it here proves to be a great expository convenience, and I am a lazy expositor. So I hope for the reader's indulgence; given which, (3.1) provides us with half of the "facts" we shall need to get started in developing higher-order electromagnetic theory.

In other words, we define or postulate the higher-order *time-like* invariant of physics to be τ and seek no further justification for this choice. Why should one have any confidence in this? The instrumentalist answer is: because we have an instrument that can *measure* this "proper time" τ. The instrument is, of course, the single clock we have alluded to, the one in an arbitrary state of motion and present at each of the two events bounding the interval $d\tau$. Its readings are irrefragable fact, agreed on by all observers (hence "invariant" under aliasing transformations)—unlike the non-invariant t, which depends on the state of motion of a plurality of co-moving clocks. Still, once we adopt the Einstein convention (discussed in all the relativity texts, equivalent to setting distant co-moving clocks by light signals of speed extrapolated to infinity—see Appendix), t becomes a useful parameter for linking up any subsequent higher-order refinements of our first-order physics—for, as will be discussed later in more detail, its differential (unlike that of proper time) is exact, hence best suited to "coordinate" representations. In a given inertial system such a t specifies the distant simultaneity on which depends Newtonian physics, including Newton's third law, *etc.*; and t is also the timelike parameter appearing in all classical theories of electromagnetism. The last three chapters of this book will be devoted to developing a slightly different kind of "t"—a "collective" variant of frame time that will benefit from the virtues of frame time for many-body description while eliminating many of its restrictions and allowing it to acquire invariance properties of its own.

Without apology, then, we shall identify the t of Eq. (3.1) with the t of our previous chapters, and shall assume that τ in (3.1) refers to the invariant proper time of the (point) *field detector*. That is, in the context of higher-order electromagnetic field theory, we consider $\tau = \tau_d$, but here and elsewhere may sometimes omit the subscript "d" denoting the detector. The latter is located at position $\vec{r} = \vec{r}(t)$, measured with respect to a given inertial system S, and is in an arbitrary state of motion relative to S specified by

$$\vec{v}_d = \vec{v}_d(t) = \frac{d\vec{r}}{dt} = \frac{d\tau}{dt}\frac{d\vec{r}}{d\tau}. \tag{3.2}$$

From (3.1) we have in general

$$\frac{d\tau}{dt} = \sqrt{1 - (dr/dt)^2/c^2} = \sqrt{1 - v^2/c^2} \equiv \frac{1}{\gamma}, \tag{3.3a}$$

or in particular for the proper time of the field detector clock-particle

$$\frac{dt}{d\tau_d} = \gamma_d = \frac{1}{\sqrt{1 - v_d^2/c^2}}. \tag{3.3b}$$

One can also define a "proper velocity" of the field detector,

$$\vec{V}_d \equiv \frac{d\vec{r}}{d\tau_d} = \frac{dt}{d\tau_d}\frac{d\vec{r}}{dt} = \gamma_d \vec{v}_d, \tag{3.4}$$

relative to S, and can obtain from (3.1) a useful general operator relationship

$$\frac{d}{d\tau} = \frac{dt}{d\tau}\frac{d}{dt} = \gamma\frac{d}{dt}. \tag{3.5}$$

The proper velocity in (3.4) is "invariant" in the sense that all observers in whatever state of motion will agree on its numerical value, provided they also measure the event separation $d\vec{r}$ in some agreed inertial system S. Unless $d\vec{r}$ is attached to a material object, it does not, of course, possess invariance in the sense that different inertial observers assign it equal numerical values with reference to their own frames ($\vec{v}_d$ and $\vec{V}_d$ both being frame dependent for the description of pairs of independent physical events).

Pitfalls of analogy: We pause to remark in connection with (3.1) that Einstein took another closely-related leap of genius, with disastrous consequences. He and Minkowski jumped to the conclusion, on the basis of "spacetime symmetry," that not only was $d\tau$ the timelike invariant of physics, but that the analogous space-dimensioned interval,

$$d\sigma = \sqrt{-c^2 d\tau^2} = icd\tau = invariant \ (?), \qquad (3.6)$$

was the spacelike invariant of physics. To an instrumentalist, this has no plausibility whatever, because there is no invariant way to measure such a quantity, no "space clock" to give it operational meaning. Its invention (out of whole cloth of the mind, without encouragement from any objective aspect of the physical world) is strictly an artefact of spacetime symmetry—an alleged "symmetry" itself created out of nothing but a parametric deficiency of Maxwell's equations—a symmetry which we have seen in Chapter 2 to be physically untenable where adequate account is taken of Faraday's observations of induction. Intelligent beings from outer space, unacquainted with Maxwell's equations, would be baffled by the concept of spacetime symmetry except as a possible aspect of religion, like the three-fold symmetry of the Trinity. In short, Eq. (3.6) asserts a physical symmetry that is not merely imaginary but in all likelihood counterfactual. Let us try to specify the trouble a little more closely.

Operationalist critique: Although I am no philosopher, let me speak here for the "discredited" viewpoint associated with what has been called *operationalism*. I have been able to discern one and only one way, consistently with the ideology of SRT, to give an operational definition of the spacelike interval defined by Eq. (3.6), and this is to make a Lorentz transformation to the inertial system in which $\delta t = 0$; for in that uniquely preferred inertial system $d\sigma = \sqrt{\delta r^2 - c^2 \delta t^2}$ becomes equal to δr. That can be measured, using as "instrument" a meter stick, so in this case $d\sigma$ is not only a real number but a measurable quantity. However, by this very token the operationalist must disavow $d\sigma$ as a physical invariant—for the fact that only a single preferred reference system exists in which $d\sigma$ can be operationally defined disqualifies that particular mathematical applicant for candidacy as physics. That is, the *operations* involved in the definition of any *physical invariant* should conform to a relativity principle: they should themselves be invariant, not dependent on frame of reference. That criterion is met by the on-trajectory invariant $d\tau$, but not by the would-be trajectory-linking "invariant" $d\sigma$. The single clock's usefulness for measuring the true physical invariant $d\tau$ holds consistently in every inertial system. The meter stick's usefulness for measuring the alleged "invariant" $d\sigma$ holds only in that preferred system in which it is at rest.

The operationalist sees such an alleged $d\sigma$-invariance as nonsense. Physical descriptive primacy belongs not to the mental creation $d\sigma$ but to the meter stick itself. That is ultimately what "invariance," as well as *measurement* (which itself needs to be an invariant concept in order to effectuate a relativity principle), is about. This suggests that there is no viable alternative to using δr itself as the spacelike invariant of physics. It is spacelike and it is measurable. Would not the existence of another spacelike, but non-measurable, invariant be something considerably worse than a redundancy? If the much put-upon *Herrgott* made δr, why in the name of all vain labors would He also make $d\sigma$? The idealized meter stick (in differing states of motion) exists invariantly in every frame. What it measures, length, exists (why not invariantly?) in every frame. Length is the very embodiment—the *sine qua non*—of the "spacelike." Try thinking of space without it. Go on—boggle your mind ... That's precisely what Einstein is asking you to do. The Einsteinians have been down in that boggle so long it looks like up to them.

Defence of operationalism: Why should one employ operational criteria in winnowing physical theories? Consider the case of $d\sigma = \sqrt{\delta r^2 - c^2 \delta t^2}$ for $\delta r > c\delta t$, so that $d\sigma$ is real. Given input numerical values of $\delta r, \delta t$, SRT predicts by paper and pencil calculation some real number $d\sigma$. In principle the input numerical values are obtained by measurement. These number inputs to the theory represent the outcomes of measurement operations we can perform with real instruments. The operationalist viewpoint is that the corresponding output $d\sigma$, *being also a real number*, should also have an operational counterpart, a measurement conceivably performed with some instrument that might be quantified by that number. In other words, a sort of logical economy is implicit: It is displeasing to have real numbers floating around in a theory with nothing measurable "out there" to attach them to. Imaginary or complex numbers are another matter, but real numbers need real referents. Any real number input to or output from a physical theory ought to pay its freight by possessing a referent in the realm of real-number *measurement*.

Bridgman's operationalism was allegedly defeated by its inability to find such a referent for the quantum-mechanical wave function. That, as I understand it, is the cause and full extent of its "discrediting." But the wave function Ψ is not a real number. When you connect it to reality *via* $\Psi * \Psi$, it submits to a probability interpretation, hence becomes measurable. Real numbers are *c*-numbers in Dirac's terminology. We need to develop greater respect for *c*-numbers. Both their presence and their absence in a theory are crucial regulators of the theory's relationship to reality. They are in fact its *only link* to reality. If I may differ from Bridgman on one point, "paper and pencil operations" detached from observability (which he introduced in desperation occasioned by the wave function issue) are exactly what must be avoided. They are the whole trouble. Uncontrolled paper and pencil operations permit the building of unlimited cloud castles of inference. Operationalism serves as a sobering antidote to this potential frivolity and its attendant trivializing of physics. (I judge febrile, over-imaginative mathematizing to be trivializing ... and the instinctive ability to recognize this to be one of the marks of a physicist.) The "invariant" $d\sigma$ exists only as a paper and pencil operation. That's what's wrong with it. Correspondingly, the strength of $d\tau$ lies in its measurability: there is an instrument that will do the job without reliance on chains of inference. That guarantees the existence of an aspect of reality corresponding to $d\tau$ and legitimizes the claim of $d\tau$ to "invariance." $d\sigma$ conspicuously lacks such legitimacy.

In summary: concerning the invariants of kinematics, the physical distinction between (3.1) and (3.6) is that (3.1) asserts an *instrumentally-measurable*, objectively real relationship between two event points on the trajectory of a single particle, whereas (3.6) asserts a corresponding *not-instrumentally-measurable* rela-

tionship between event points on two different particle trajectories, such that a light signal cannot connect those two events.
When you think about the latter "world structural" proposition,
you will see that it is truly spooky: Here is an alleged descriptor
of "reality," a number inaccessible to instrumental measurement
and accessible only to calculation, such that even a light signal
cannot check on it; yet we hold it, as an article of faith, to be more
real than our yardstick, more real than the stone that Samuel
Johnson kicked ... because the yardstick and the stone (saith Einstein, forsooth) are non-invariant, whereas the calculated interval
(3.6) is claimed to be invariant. The claim rests on no firmer basis
than beauty (a certain imagined symmetry—and an erroneous
one at that ...). A calculated number is thus more real than the
stone that bruises our foot. And Einstein in later life disapproved
of spooky things! Remember: Invariance is not something to kid
about ... it is to be used sparingly, with reverence, to *describe* what
is *really* "out there" in nature. Should not "really" mean verifiably? Did not Dr. Johnson have the right idea?

So, we are free (and well-advised) to disregard (3.6) and all
the evil empire built upon it, and to postulate a different higher-
order spacelike invariant of physics. What shall we choose? All
the world is at our disposal. The simplest possibility, already adduced, is that *object length* (measured as Euclidean separation distance of endpoints) is the higher-order invariant. Here is the object. It is part of our experience—that which it is theory's job to
describe. The object is thus as "real" as our experience, which is
to say as real as we are. If it is real, there should be in our theory
an invariant to match it. Unless there is good reason to seek
something more sophisticated, it makes sense to try that first.
Loose cannons, loose realities adrift without invariant theoretical
counterparts—like loose invariants without counterparts in reality—are surely to be avoided in sound theory. The operationalist
or instrumentalist will point to a meter stick as a perfectly good
instrument for measuring length, and I know of nothing better,
although a yardstick is just as good. We have educated our youth
to higher things, to be sure, but the scorn of youth is no harder to
bear than that of age and century-long tradition. (All
right ... meter sticks are affected by temperature and light waves
aren't, so no modern experimentalist would dream of using a meter stick for accurate measurement of object length. But he would
use *some* instrument ... and my "meter stick" is in any case a *Ge-*

danken one immune to mundane perturbations. The secret of doing physics lies in the finding of harmless idealizations—those
that reveal more than they conceal. There is no formula for it. It is
an art ... but also a matter of taste, guided by experience. For instance, I happen to like rigid bodies, as will be discussed in Chapter 4—they suit my taste. Relativists abhor them and cannot co-
exist in the same "world" with them.)

Why, indeed, should we seek anything better than Euclidean
length for our invariant? At this point vague rumblings are apt to
be heard from some quasi-physicists about the Michelson-Morley
experiment[3.1]—to the effect that length has to contract to prevent
fringes from shifting, so length cannot be a higher-order invariant. The M-M experiment, to which we have already alluded,
merely extends to higher order the first-order observations of
Mascart[2.6] and others of the nineteenth century showing the
physical validity of a relativity principle. Any mathematics that
supports a relativity principle is supported by M-M. And we saw
that Hertz's invariant mathematics did a perfectly good job of that
at first order. In order to accommodate M-M automatically at
higher order, all we have to do is find a higher-order *invariant*
counterpart of Hertz's mathematics. Indeed, we can accomplish
that with almost any choice of spacelike invariant ... but the
choice of object length is surely the first one to try, and we shall
find no need to go beyond it. The proof will be in the pudding.

In summary, then, the ratiocinations of this section have led
us to postulate the invariants of kinematics to be:

$$\textbf{Timelike}: d\tau = \sqrt{dt^2 - dr^2/c^2} = invariant$$
$$\textbf{Spacelike}: \delta r = invariant.$$

(3.7)

These will be our postulates throughout this book—which means
we accord them in effect a factlike status. In (3.7) we use a notation dr to denote spatial separation of two successive events (of
timelike separation) on the trajectory of a single particle, and distinguish this from δr, which denotes spacelike separation of
events marking, *e.g.*, the two ends of an extended structure, a meter stick. [To be entirely consistent, we should be able to speak of
the "distant simultaneity" of such events. The reader must take
this on trust for now, as we defer to Chapter 6 our investigation of
the absoluteness of simultaneity. Here it will suffice that
$\delta r = invariant$ be understood to mean that extended structures
undergo no metric (measurable dimensional) changes, such as

Lorentz contraction, as a result of relative motion or of changes of environment, such as gravity potential. In the relativity language, which we use sparingly in this book, d in (3.7) denotes an on-worldline event separation, whereas δ denotes a separation of events on two different worldlines.] Of course we have not said anything yet about what the physical "inertial transformations" are, mathematically, at higher order. (We consider them to be Galilean at first order.) Nor have we examined higher-order kinematics. For the moment we take a carefree attitude toward those problem areas, to be treated later, since electromagnetism is our immediate concern. Einstein put electromagnetism ahead of mechanics—making the latter conform to the former. This is a distorted judgment, but let us accept it for the moment and proceed.

3.2 Neo-Hertzian field equations

The first-order Hertzian field equations, (2.4), were patterned faithfully on the corresponding Maxwell equations, (1.1). The only difference, apart from source term adjustments (resulting from detector motion), was that Maxwell's $\partial/\partial t$ was replaced everywhere by d/dt, where (in both cases) t is the frame time of an inertial system. To proceed beyond this first-order approximation, it is evident that we must retain the "total" aspect of Hertz's total time derivative and make a further replacement in his first-order equations of non-invariant frame time t by the higher-order *invariant* proper time τ of something. Proper time of what? Of the field detector, of course. We keep coming back to that … "fields" being what field theory is about, and "fields" being what field detectors detect. Recall that we idealize the field detector as a mathematical point sufficiently massive to possess a trajectory, so there is no ambiguity about measurement of this τ. It is the time displayed by a co-moving idealized classical "clock-particle." Such a clock or pocket-watch is just one more of the co-moving instruments incorporated in the multi-purpose instrument composite we have idealized as the "point detector." The motion need not be inertial but can be represented by any function of frame time t, which by (3.7) is related to τ by

$$\tau = \int \sqrt{1 - v_d^2/c^2}\, dt, \tag{3.8a}$$

where $v_d = v_d(t) = dr(t)/dt$ is an arbitrary smooth function descriptive of detector velocity measured in an inertial frame. For $v = v_d = const.$, this reduces to

$$\tau = \frac{t}{\gamma}, \tag{3.8b}$$

where γ is given by (3.3a).

As just indicated, we carry forward to higher approximations the *covering theory* theme by formally replacing the non-invariant t wherever it appears in the Hertzian field equations (2.4) by the higher-order invariant parameter $\tau = \tau_d$, the proper time of the field detector. Thus the higher-order invariant field equations, here termed *neo-Hertzian*, are postulated to be

$$\vec{\nabla} \times \vec{B} - \frac{1}{c}\frac{d\vec{E}}{d\tau} = \frac{4\pi}{c}\vec{j}_m \tag{3.9a}$$

$$\vec{\nabla} \times \vec{E} = -\frac{1}{c}\frac{d\vec{B}}{d\tau} \tag{3.9b}$$

$$\vec{\nabla} \cdot \vec{B} = 0 \tag{3.9c}$$

$$\vec{\nabla} \cdot \vec{E} = 4\pi\rho . \tag{3.9d}$$

This is a covering theory of the Hertzian first-order equations (2.4), because τ differs from t only at second order. Hence, as detector velocity slows relative to our inertial system S, τ becomes indistinguishable from t and (3.9) goes identically into (2.4) as a "covered" case. The latter is itself a covering theory of Maxwell's equations, so all the observational-agreement credits of the accepted school-taught electromagnetic theory automatically accrue to the neo-Hertzian variant.

The length-invariance *Ansatz*, Eq. (3.7), carried over from first order, ensures that the grad operator $\vec{\nabla}$ remains invariant at higher orders, as at first order [Eq. (1.3a)]. It also ensures that our small detector volume does not change its dimensions in any observer's view; hence, the enclosed-charge count stays the same and Eq. (2.7) remains true at higher orders:

$$\rho'(\vec{r}',\tau') = \rho(\vec{r},\tau). \tag{3.10}$$

Similarly for Eq. (2.8),

$$\vec{j}_s(\vec{r}',\tau') = \vec{j}_s(\vec{r},\tau) - \rho(\vec{r},\tau)\vec{V} \tag{3.11}$$

holds, where in formal analogy to Eq. (3.4) we have from (3.1) a proper relative velocity of inertial frame S' with respect to S,

$$\vec{V} = \frac{d\vec{r}}{d\tau} = \frac{dt}{d\tau}\frac{d\vec{r}}{dt} = \frac{\vec{v}}{\sqrt{1-(v/c)^2}} = \gamma\vec{v}. \qquad (3.12)$$

Here the proper time τ is that of any point at rest in inertial frame S'. Evidently in this special case τ can be identified with the frame-time t' of S', and $\vec{v}$ (a constant) is the frame-time velocity of S' measured by instruments at rest in S.

In order to proceed, we need to sketch a higher-order analog of the Galilean velocity composition law, Eq. (2.1). Consider our point detector to be instantaneously located at position $\vec{r}_d'$ in inertial frame S', which moves at uniform velocity $\vec{v}$ relative to S. At all orders, its rigorously exact location in S is

$$\vec{r}_d = \vec{r}_d' + \vec{v}t, \qquad (3.13)$$

where t is the frame-time of S. Transposing and differentiating with respect to t, we find

$$\frac{d\vec{r}_d'}{dt} = \frac{d\vec{r}_d}{dt} - \vec{v} \quad \rightarrow \quad \vec{v}_d' = \vec{v}_d - \vec{v}, \qquad (3.14)$$

the implication being valid only at first order, where $t'=t$, in agreement with the Galilean transformation (1.2). Thus we confirm our first-order velocity addition result, (2.1). However, if instead we differentiate (3.13) with respect to detector proper time τ, we get

$$\frac{d\vec{r}_d'}{d\tau} = \frac{d\vec{r}_d}{d\tau} - \vec{v}\frac{dt}{d\tau} = \frac{d\vec{r}_d}{d\tau} - \vec{v}\gamma, \qquad (3.15)$$

or, with (3.4) and (3.12),

$$\vec{V}_d' = \vec{V}_d - \vec{V}, \qquad (3.16)$$

which is the proper-time form of our desired higher-order velocity composition law. In the above calculations we have dropped the notational distinction between δr and dr, since it is evident that spatial-increment lengths are meant in each case with reference to a particular inertial system. Finally, nothing in our discussion limits the detector to immobility in S'. The result (3.16) therefore remains valid if arbitrary detector motion $\vec{V}_d(\tau)$ in S is described in S' by $\vec{V}_d'(\tau')$ for instantaneous velocities. (Keep in mind that $\vec{V}$ is a *constant* for inertial motion.)

Applying (3.12) in S and in S' we can write out (3.16) in terms of frame-time velocities:

$$\frac{\vec{v}_d'}{\sqrt{1-\left(v_d'/c\right)^2}} = \frac{\vec{v}_d}{\sqrt{1-\left(v_d/c\right)^2}} - \frac{\vec{v}}{\sqrt{1-\left(v/c\right)^2}}. \qquad (3.17)$$

In the simplest case of collinear velocities, say, a particle moving with speed $w = v_d'/c$ with respect to S', speed $u = v_d/c$ with respect to S, and S' moving with speed $z = v/c$ with respect to S, all moving along the same line, (3.17) becomes

$$\frac{w}{\sqrt{1-w^2}} = \frac{u}{\sqrt{1-u^2}} - \frac{z}{\sqrt{1-z^2}} \qquad (3.18)$$

for w, u, z each restricted to the open interval between -1 and 1, to avoid non-physical singularities. [This means, as in SRT, that all frame-time velocities are limited in magnitude to c as an unreachable upper limit (*pace* tachyons).] The quadratic (3.18) can be solved for w to yield

$$w = \pm\sqrt{N/D}, \qquad (3.19)$$

where

$$N = 2uz\sqrt{1-u^2}\sqrt{1-z^2} + 2u^2z^2 - u^2 - z^2$$
$$D = 2uz\sqrt{1-u^2}\sqrt{1-z^2} + u^2z^2 - 1.$$

When this w-function is plotted, the choice of roots becomes obvious. Since the non-collinear case is still more complicated, it is apparent that this law of velocity composition, based on proper times, is not simply expressible in terms of frame-time velocities. Moreover, (3.17) does not describe the most general case of relative velocity of two arbitrarily-moving particles viewed in different inertial frames. It treats only the case of a single particle as viewed in two different inertial frames ... but it does so by means of two proper times—that of the particle moving arbitrarily with instantaneous speed u in S and that of any point at rest in S' (the latter proper time being equivalent to the frame time t' in S'). We defer further discussion of topics related to kinematics to later chapters, as this would distract us here. For the present all we need to retain is that analysis in terms of proper times entails great complexity.

Because our velocity composition laws, (2.1) at first order and (3.16) at higher orders (actually, simple "addition" laws), conveniently have identically the same form, all formal aspects of the invariance proof for Hertzian field equations hold for the neo-Hertzian ones. The invariance proof for the higher-order case therefore need not be repeated. We need only pause to note that the higher-order "inertial" transformations under which this invariance holds, which might be termed "neo-Galilean," are

$$\vec{r}' = \vec{r} - \vec{V}\tau, \quad \tau' = \tau .$$ (3.20)

Here $\tau' = \tau$ has the trivial meaning that the same clock or particle shows the same elapsed times as viewed by differently-moving observers. The general relation

$$\vec{V}d\tau = \gamma \vec{v}(dt/\gamma) = \vec{v}dt$$ (3.21a)

[which follows from (3.12) and (3.3a)] can be integrated in the case of constant $\vec{v}$ (implying constant $\vec{V}$) to

$$\vec{V}\tau = \vec{v}t .$$ (3.21b)

Hence the spatial part of the neo-Galilean transformation (3.20) can be written in traditional Galilean form,

$$\vec{r}' = \vec{r} - \vec{v}t ,$$ (3.22)

while the time part can be written [by identifying τ' as t' for the case of the detector at rest in S' and applying (3.3a) with the recognition that $\vec{v}$ is constant] as

$$t' = t/\gamma$$ (3.23)

in frame-time (not explicitly invariant) form. This expresses "relativistic" time dilation without the clock-phase dependence on distance claimed by SRT. Needless to say, it is the time dilation aspect that is confirmed by experiment. (It may be added that Franco Selleri[3.2] has argued in numerous publications for (3.23)—which allows for absolute simultaneity—as the physically correct representation of time dilation. However, he also insists on the Lorentz contraction of spatial quantities ... and on this we disagree.)

3.3 Neo-Hertzian wave equation

The formal similarity of the higher-order field equations (3.9) to the Hertzian ones, (2.4), allows us to set down immediately the neo-Hertzian wave equation by analogy with (2.18),

$$\nabla^2 \vec{E} - \frac{1}{c^2}\frac{d^2\vec{E}}{d\tau^2} = 0 .$$ (3.24)

This equation has been treated previously.[2.11, 3.3] Considering first the case that $\vec{v}_d$ (hence γ_d) is constant, we shall give two different derivations of its solution.

Solution using invariant forms. The invariances $\vec{E}' = \vec{E}$ of Eq. (2.6) and $c' = c$ of Eq. (2.13) may be assumed valid at higher orders. Hence the wave equation (3.24) is of manifestly invariant form. It should therefore be capable of solution by means employing only invariant quantities. Calling to mind the first-order wave equation (2.18), namely,

$$\nabla^2 \vec{E} - \frac{1}{c^2} \frac{d^2 \vec{E}}{dt^2} = 0,$$

and its solution (2.24), namely,

$$u = \frac{\omega}{k} = \pm c + \frac{\vec{k}}{k} \cdot \vec{v}_d,$$

we have only to introduce higher-order counterparts u^*, ω^* of u, ω and to recognize [for example by comparing (3.12) with (2.8)] that $\vec{V}$ is the higher-order counterpart of $\vec{v}$, to set down at once by formal similarity the solution of (3.24) as

$$u^* = \frac{\omega^*}{k} = \pm c + \frac{\vec{k}}{k} \cdot \vec{V}_d. \tag{3.25}$$

Here we have taken account of $\vec{k}^* = \vec{k}$, which reflects the fact that spatial vector quantities such as $\vec{k}$ behave in a classical way (in consequence of length invariance, implying Euclidean geometry). In order to "translate" (3.25) into a more useful frame-dependent form, we recognize that

$$\omega^* \tau = \omega t; \tag{3.26}$$

that is, this dimensionless product represents the same pure number, whether expressed in terms of invariant or frame-dependent quantities (there being no juncture in the smooth transition from lower- to higher-order description at which a numerical discontinuity can occur). Then

$$\frac{\omega^*}{\omega} = \frac{t}{\tau} = \frac{dt}{d\tau_d} = \gamma_d \tag{3.27}$$

from (3.3b). Hence

$$u^* = \frac{\omega^*}{k} = \frac{\omega \gamma_d}{k} = u \gamma_d, \tag{3.28}$$

and Eq. (3.25) can be written with the aid of (3.4) as

$$u^* = u \gamma_d = \frac{\omega \gamma_d}{k} = \pm c + \frac{\vec{k}}{k} \cdot \gamma_d \vec{v}_d. \tag{3.29}$$

Finally, dividing through by γ_d, we arrive at

$$u = \frac{\omega}{k} = \pm \sqrt{c^2 - v_d^2} + \frac{\vec{k}}{k} \cdot \vec{v}_d, \tag{3.30}$$

which is our higher-order wave equation solution expressed in frame-measurable form, for easiest comparison with observation. We shall make much use of this result in what follows.

Solution using frame-dependent forms. This treatment is more tedious but also more straightforward. As above, we simplify by considering $\vec{v}_d$ (hence γ_d) to be a constant. Following our method of solving the Hertzian equation, we seek a solution of the form $\vec{E} = \vec{E}(p)$, where

$$p = \vec{k} \cdot \vec{r} - \omega t ,\qquad (3.31)$$

the same as (2.19). Then, as in Eq. (2.20), we have

$$\nabla^2 \vec{E}(p) = \left(\frac{\partial^2}{\partial x^2} + \frac{\partial^2}{\partial y^2} + \frac{\partial^2}{\partial z^2} \right) \vec{E}(p) = k^2 \vec{E}''(p) .\qquad (3.32)$$

Applying (3.5) with $\tau = \tau_d$ and (1.9), we find for constant γ_d

$$\frac{d^2}{d\tau^2} \vec{E}(p) = \gamma_d^2 \left(\frac{\partial}{\partial t} + \vec{v}_d \cdot \vec{\nabla} \right)^2 \vec{E}(p)$$

$$= \gamma_d^2 \left[\frac{\partial^2}{\partial t^2} + 2 \frac{\partial}{\partial t} \left(\vec{v}_d \cdot \vec{\nabla} \right) + \left(\vec{v} \cdot \vec{\nabla} \right)^2 \right] \vec{E}(p)\qquad (3.33)$$

$$= \gamma_d^2 \left[\omega^2 - 2\omega \left(\vec{v}_d \cdot \vec{k} \right) + \left(\vec{v}_d \cdot \vec{k} \right)^2 \right] \vec{E}'' = \gamma_d^2 \left[\omega - \left(v_d \cdot \vec{k} \right) \right]^2 \vec{E}'' .$$

From (3.24), (3.32) and (3.33) it follows that

$$\left\{ k^2 - \frac{\gamma_d^2}{c^2} \left[\omega - \left(\vec{v}_d \cdot \vec{k} \right) \right]^2 \right\} \vec{E}''(p) = 0 .\qquad (3.34)$$

From the vanishing of the coefficient of $\vec{E}''$ we have

$$ck = \gamma_d \left| \omega - \vec{v}_d \cdot \vec{k} \right| ,\qquad (3.35)$$

or, dividing by $k\gamma_d$ and defining a frame-time phase velocity u as before, we get

$$u = \frac{\omega}{k} = \pm \sqrt{c^2 - v_d^2} + \frac{\vec{k}}{k} \cdot \vec{v}_d ,\qquad (3.36)$$

in agreement with our previous solution (3.30).

Finally, it is of possible interest to remove the above restriction to constant $\vec{v}_d$.

Solution for arbitrary $\vec{v}_d = \vec{v}_d(t)$. In this more general case[3.4] we look for a d'Alembertian solution of the form $\vec{E} = \vec{E}(p)$ where

$$p = \vec{k} \cdot \vec{r} - f(t) ,\qquad (3.37)$$

$\vec{k}$ being constant as before and f an arbitrary function that generalizes the previous ωt. The spatial part, (3.32), holds as before.
The time part is

$$\frac{d^2}{d\tau_d^2}\vec{E}(p) = \gamma_d \frac{d}{dt}\gamma_d \frac{d}{dt}\vec{E}(p). \tag{3.38}$$

In order to evaluate this, we need to know dp/dt. From (1.9)

$$\frac{dp}{dt} = \dot{p} = \left(\frac{\partial}{\partial t} + \vec{v}_d \cdot \vec{\nabla}\right)\left(xk_x + yk_y + zk_z - f(t)\right)$$

$$= -\dot{f} + v_{dx}k_x + v_{dy}k_y + v_{dz}k_z = -\dot{f} + \vec{v}_d \cdot \vec{k}, \tag{3.39}$$

where $\dot{f} \equiv df/dt$. From (3.38) we obtain

$$\gamma_d \frac{d}{dt}\left(\gamma_d \frac{d}{dt}\vec{E}(p)\right) = \gamma_d \dot{\gamma}_d \dot{p}\vec{E}' + \gamma_d^2 \ddot{p}\vec{E}' + \gamma_d^2 \dot{p}^2 \vec{E}'', \tag{3.40}$$

with $\dot{\gamma}_d \equiv d\gamma_d/dt$. Use of this and (3.32) in the wave equation
(3.24) yields

$$\left(k^2 - \frac{\gamma_d^2 \dot{p}^2}{c^2}\right)\vec{E}'' - \frac{\gamma_d^2}{c^2}\left(\ddot{p} + \frac{\dot{\gamma}_d \dot{p}}{\gamma_d}\right)\vec{E}' = 0. \tag{3.41}$$

In order for a d'Alembertian solution to exist it is necessary that
the coefficients of both $\vec{E}'$ and $\vec{E}''$ vanish. For $\vec{E}'$ this means

$$\frac{d^2 p}{dt^2} = -\frac{1}{\gamma_d}\left(\frac{d\gamma_d}{dt}\right)\frac{dp}{dt}.$$

Let $y \equiv dp/dt$. Then

$$\frac{dy}{dt} = -\frac{1}{\gamma_d}\left(\frac{d\gamma_d}{dt}\right)y \quad \rightarrow \quad \frac{dy}{y} = -\frac{d\gamma_d}{\gamma_d}. \tag{3.42}$$

The solution of this is $y = b/\gamma_d$, with b an integration constant.
With (3.39) this yields

$$y = \frac{dp}{dt} = -\dot{f} + \vec{v}_d \cdot \vec{k} = \frac{b}{\gamma_d}. \tag{3.43}$$

The vanishing of the coefficient of $\vec{E}''$ in (3.41) implies that

$$\frac{ck}{\gamma_d} = \left|\frac{dp}{dt}\right| = \left|-\dot{f} + \vec{v}_d \cdot \vec{k}\right|. \tag{3.44}$$

Taking the absolute value of (3.43) and comparing with (3.44), we
see that $|b| = ck$, i.e.,

$$b = \pm ck. \tag{3.45}$$

We assume this condition, which assures simultaneous vanishing of the coefficients of $\vec{E}', \vec{E}''$ in (3.41) to be satisfied. Eq. (3.43) then implies

$$\dot{f} = \pm\frac{ck}{\gamma_d} + \vec{v}_d \cdot \vec{k} , \qquad (3.46)$$

whence

$$f(t) = \int_0^t \dot{f}dt = \int_0^t \left(\pm\frac{ck}{\gamma_d} + \vec{v}_d \cdot \vec{k} \right) dt . \qquad (3.47)$$

(Here it will be recalled that $\vec{v}_d$ and γ_d are functions of t.) The d'Alembertian wave function argument p can always be written in the form $\vec{k}\cdot\vec{r} - \omega t = \vec{k}\cdot\vec{r} - ukt$, which may be considered to define the phase velocity u. Comparing with (3.37), we see that $ukt = f(t)$; so

$$u = \frac{f(t)}{kt} = \pm\frac{1}{t}\int_0^t \sqrt{c^2 - v_d^2}\,dt + \frac{1}{kt}\int_0^t \vec{v}_d \cdot \vec{k}dt$$

$$= \pm\left\langle \sqrt{c^2 - v_d^2} \right\rangle_{av} + \left\langle \vec{v}_d \cdot \frac{\vec{k}}{k} \right\rangle_{av} , \qquad (3.48)$$

where $\langle\ \rangle_{av}$ denotes a time average over the interval zero to t.

Eq. (3.48) is our desired generalization of the previous results (3.30) and (3.36), which were obtained for the special case of $\vec{v}_d$ constant. It will be observed that the form of the solution is not affected, beyond introduction of a time-averaging process over an interval of length t. What is this t? Here we need help from the physics. Two likely candidates suggest themselves: (1) It is the propagation time of the photon from emission to absorption. (2) It is the absorption time interval. The first of these seems ruled out by common sense, as some starlight (presumably, on the customary causal model) has been propagating since long before the Earth came into existence, and it seems infeasible to "average" over the motions of an absorber not yet in existence during part of the alleged averaging interval. More plausible is alternative (2), since it asks us to average only during an absorption process we know must be occurring because it constitutes the "field detection" our field theory is concerned with. Adopting this view, we can say that as a practical matter the interval zero to t can be considered so short (of the order of 10^{-18} second for single-photon absorption, or flash duration for a group of photons) as to be practically instantaneous. Thus the "averaging" is in general an unnecessary refinement, and in treating problems of wave (radiation) detection we can view $\vec{v}_d$ in our formulas as having a constant instantaneous value associated with the moment of detection. However, this question should not be considered settled on the basis of mere ratiocination. It deserves to be kept open.

The upshot is that all three of the above methods of solving the neo-Hertzian wave equation tentatively agree on a solution having phase velocity (propagation speed)

$$u = \frac{\omega}{k} = \pm\sqrt{c^2 - v_d^2} + \frac{\vec{k}}{k} \cdot \vec{v}_d . \tag{3.49}$$

Referring to Eq. (3.31), we see that the most general d'Alembertian solution of that equation is consequently

$$\vec{E} = \vec{E}_1\left(\vec{k}\cdot\vec{r} + \left(k\sqrt{c^2 - v_d^2} - \vec{k}\cdot\vec{v}_d\right)t\right)$$
$$+ \vec{E}_2\left(\vec{k}\cdot\vec{r} - \left(k\sqrt{c^2 - v_d^2} + \vec{k}\cdot\vec{v}_d\right)t\right), \tag{3.50}$$

where $\vec{E}_1, \vec{E}_2$ are arbitrary vector functions. Thus the neo-Hertzian light speed u is not a constant except in the special (Maxwellian) case of a stationary detector, $v_d = 0$.

Our discussion of Potier's principle in Chapter 2, establishing the unobservability (by ordinary laboratory techniques of interferometry, *etc.*) of an additive term of type $\vec{k}\cdot\vec{v}_d$ in the phase velocity, of course applies here as well. Because of this unobservability, little attention need be paid to the apparently "acausal" aspects of the description of light "propagation" given by (3.50). Still, there are many who will be disturbed by the implication that the emitter "knows" enough about the future absorption event to regulate photon speed. It is worth pausing to note once again that this confronts us directly with the quantum nature of basic electromagnetic processes—at least those observed in the far zone ("radiative"). Quantum mechanics is fundamentally acausal by the very nature of its pedigree, being based upon a formal Correspondence with a classical instant-action-at-a-distance form of mechanics.

Moreover "propagation," as we have repeatedly noted, is Everyman's paradise of inference—being as far from direct operational verification as ever an "ether wind" was. Einstein chose to think about propagation in exactly the same plodding, causal way Maxwell did, but without the concrete justification of Maxwell's picture of contiguous contacts within an ether to flesh-out a physical picture of *what* was doing the propagating. So, Einstein has left us with mathematical vectors "propagating" (straight out of our minds) through physical space. Why such fancy-free mathematical creations should be fettered by further imaginings of etherless contiguous causal evolution, as if successive ghostly "contacts" operated upon "vectors" across space, is hard to trace to anything but cultural inertia of physical theorists or lack of roughage in their mental diet. In fact, trying to picture what is going on during propagation is identically the same thing as trying to picture what is going on in a quantum pure state—for the simple reason that the photon propagates in a pure state. The dangers of pictorialization are notorious, since they make up one of the central discovery themes of early twentieth-century physics. Of course, Einstein didn't know about that in 1905, and never wanted to know about it. Can you blame him? He was the last causal thinker about the quantum world, except for ten million successors who carry on his tradition to this day by thinking causally about light propagation ... and being cocksure about it, at that, by virtue of their unwavering allegiance to Maxwell's equations.

Meanwhile, are there published premonitions of acausal formulations of electromagnetic theory, including non-standard views of "propagation"? Indeed, many such. Right off the bat we notice that Maxwell's equations themselves specify no inherent preference between advanced and retarded descriptions of "propagation." It is we, the users, who superpose that gratuitously by selecting retarded and discarding advanced solutions of the wave equation. That's our privilege, to be sure, but let's not be cocksure about it! As Oliver Cromwell wrote in 1650 to the General Assembly of the Church of Scotland, "I beseech you, in the bowels of Christ, think it possible you may be mistaken." A light-sphere shrinking in acausal response to a detection event is operationally indistinguishable from a light-sphere expanding in causal response to an emission event. And, of course, light-spheres have little to do with the quantum reality, anyway—for all that Huyghens was a great man.

The ambiguity inherent in this situation allowed Wheeler-Feynman to propound their theory[3.5] of "the absorber as the mechanism of radiation." This employed a balanced combination of half-advanced and half-retarded potentials, with an assumption of perfect absorption, such that no photon sets out on its journey without being assured of an absorber to arrive at. This is another way of edging up on our present Hertzian suggestion that absorber motion can influence one-way light "propagation" speed. And, to assert the (causal) hypothesis that the universe is guaranteed to be so full of absorbers that no photon is ever orphaned (and energy is thus conserved), is surely a postulationally spendthrift way of achieving what is more efficiently accomplished by the (acausal) hypothesis that no photon is ever *emitted* without its absorption being pre-arranged. The latter requires us to feign no hypotheses about the "universe." Fokker (I lack the reference, but it can be found in Wheeler and Feynman[3.5]) said, "The sun would not radiate if it were alone in space." He had the right idea, perhaps. Another unconventional thinker about photons and their absorption was G. N. Lewis (who first proposed the name "photon" in 1926).

Finally, it might be noted that SRT itself provides one prominent opening for unconventional views of light. The timelike interval of that theory, $d\tau$, vanishes for any two event points connected by a light signal. This could be interpreted to mean that the emitting and absorbing atoms are in "virtual contact." (If I am not mistaken, this, too, was G. N. Lewis's idea and terminology.) That is, the photon ages by an amount *zero* during its "journey;" so the journey is *in some sense* of zero spatial length (if we take the photon's word for it). If we don't understand that, it may be merely because we are not photons. These matters are mentioned here only to bring out the point that spookiness in the description of light "propagation" is not a private innovation of this writer's, but is in fact something of a hardy-perennial cottage industry ... or once was, before the great SRT mental ice age set in hard.

3.4 Phase invariance

Another implication of the neo-Hertzian wave equation concerns the transformation properties of phase: The first-order invariant relation (2.6) applies also at higher orders,

$$\vec{E}'(p') = \vec{E}(p).$$ (3.51)

This implies numerical phase equality, $p' = p$; hence from (3.31)

$$\vec{k} \cdot \vec{r} - \omega t = \vec{k}' \cdot \vec{r}' - \omega' t'.$$ (3.52)

This expresses the classical principle of phase invariance. For simplicity, consider the detector at rest in inertial system S', which moves with velocity $\vec{v}$ with respect to S. Then $\vec{v}_d = \vec{v}$ and the proper time τ_d of the detector is equal to the S' frame time $t' = t/\gamma_d$, according to Eq. (3.23). Using this and Eq. (3.22) to eliminate the $(\vec{r}', t')$ variables, we obtain

$$\vec{k} \cdot \vec{r} - \omega t = \vec{k}' \cdot (\vec{r} - \vec{v}_d t) - \omega' t/\gamma_d$$

or

$$\left(\vec{k} - \vec{k}'\right) \cdot \vec{r} = \left(-\vec{k}' \cdot \vec{v}_d + \omega - \omega'/\gamma_d\right)t. \tag{3.53}$$

Since $\vec{r}$ and t may vary independently, their coefficients must vanish. Hence we obtain the higher-order generalizations of (2.28) and (2.29):

$$\text{Coeff. of } \vec{r}: \qquad \vec{k}' = \vec{k} \tag{3.54}$$

$$\text{Coeff. of } t: \qquad \omega' = \gamma_d\left(\omega - \vec{k} \cdot \vec{v}_d\right). \tag{3.55}$$

The second of these results describes the Doppler effect for source stationary in S and detector at rest in S'. The first seems to describe aberration—and to do so incorrectly, by denying that light propagation vectors are directionally affected by inertial transformations. Since such an effect of light $\vec{k}$-vector turning is a core teaching of SRT, and since the subject will turn out to have some subtleties, as well as considerable interest in its own right, we shall devote the next chapter to the topic of stellar aberration, basing our discussion there on our principal neo-Hertzian results, (3.49) and (3.54). For now, let us examine the Doppler effect.

3.5 Doppler effect

Recall that in deriving our Doppler result, Eq. (3.55), we assumed the detector to be at rest in S'. Hence the S' observer sees a motionless detector, $v'_d = 0$, and we have the Maxwellian case $u' = c$ and $\omega' = k'c = kc$, in view of $k' = k$ from (3.54). The "transmitted" frequency of the light source, $\omega = \omega_0$, measured in the rest system S of the source, then obeys

$$\omega_0 = \omega' \gamma_d^{-1} + \vec{k} \cdot \vec{v}_d, \tag{3.56}$$

from (3.55); whence

$$\omega' = \omega_0 \gamma_d \left(1 + \ell \gamma_d \frac{v_d}{c}\right)^{-1} \approx \omega_0 \left[1 - \ell \frac{v_d}{c} + \left(\ell^2 + \frac{1}{2}\right)\left(\frac{v_d}{c}\right)^2 + \cdots\right], \tag{3.57}$$

where $\ell = \left(\vec{k}/k\right) \cdot \left(\vec{v}_d/v_d\right)$ is the cosine of the angle between the light propagation direction and the detector velocity measured in the source system S, and we have used $\vec{k} \cdot \vec{v}_d = \ell k v_d = \ell v_d \omega'/c$ in (3.56). Clearly, to describe the stellar Doppler effect, it is necessary to take $\vec{v}_d = \vec{v}_{source-sink}$, the velocity of the telescope relative to the star—which is of course different for each stellar source.

  Eq. (3.57) is to be compared with a standard result of SRT for the Doppler effect [see Chapter 4, Eq. (4.2d)]; *viz.*,

$$\omega'_{SRT} = \omega_0 \gamma_d \left(1 - \ell\frac{v_d}{c}\right) \approx \omega_0 \left[1 - \ell\frac{v_d}{c} + \frac{1}{2}\left(\frac{v_d}{c}\right)^2 + \cdots\right]. \qquad (3.58)$$

This has been discussed in *Heretical Verities*.[2.11] [Since $\vec{v}_d$ is detector velocity relative to source (S' relative to S), the way of looking at it that employs source velocity relative to the detector must replace v_d by $-v_{source}$ in these formulas.] The neo-Hertzian result (3.57) agrees at first order with the SRT result (3.58), but disagrees at second order. Unfortunately, the disagreement is not easily tested. For the case $\ell = 0$ of transverse Doppler, the two formulas agree with each other and with the observationally confirmed time dilation, to second order, which is all that can be observed by methods such as those employed by Ives-Stillwell.[3.6] For the case $\ell \neq 0$ the first-order term is non-vanishing and dominates over the very small second-order term. So the older experimental methods will not serve and no existing data can settle the issue. However, frequency measurements by modern techniques have become extremely precise, so it is not out of the question that a crucial experiment could be devised to allow a choice between (3.57) and (3.58). This has not been investigated ... but it offers the possibility of an independent empirical check on other claims in this book

3.6 Chapter summary

In this chapter we have set up the most obvious form of field theory based on Hertz's theme of invariance, and have carried this one step beyond the first-order stage treated in Chapter 2 by substituting the invariant proper time of the field detector for non-invariant frame time. The resulting higher-order invariant electromagnetic theory, which includes the effects of time dilation, we have termed neo-Hertzian. On the spatial side we elected to leave well enough alone and try the simplest thing, length invariance and Euclidean geometry. In the next chapter we shall see that this succeeds in describing stellar aberration where SRT (as its supporters are blissfully unaware) fails irremediably. In this connection we shall be led to propose an astronomical observation both feasible and crucial, in the sense that it should (if there is such a thing as a crucial experiment) settle for all time the issue between invariance and covariance. Here we have prepared the way by working out in full detail, by three different (but mutually consistent) methods of calculation, the solution of the neo-Hertzian wave equation. This tells us everything that classical field theory, amended to make it invariant under inertial transformations, has to say about the propagation of light.

References for Chapter 3

[3.1]   A. A. Michelson and E. W. Morley, *Am. J. Science* **31**, 377-386 (1886); *ibid.*, **34**, 333-345 (1887).

[3.2]   F. Selleri, Found. Phys. **26**, 641-664 (1996); "Sagnac effect: end of the mystery," in *Relativity in Rotating Frames*, G. Rizzi and M. L. Ruggiero, eds. (Kluwer, Dordrecht, 2004), pp. 57-77.

[3.3]   T. E. Phipps, Jr., "Hertzian Invariant Forms of Electromagnetism," in *Advanced Electromagnetism Foundations, Theory and Applications*, T. W. Barrett and D. M. Grimes, eds. (World Scientific, Singapore, 1995).

[3.4]   T. E. Phipps, Jr., *Apeiron* **7**, 76-82 (2000).

[3.5]   J. A. Wheeler and R. P. Feynman, *Rev. Mod. Phys.* **17**, 157 (1945) and **21**, 425 (1949).

[3.6]   H. E. Ives and G. R. Stilwell, *J. Opt. Soc. Am.* **28**, 215 (1938); **31**, 368 (1941).

Having set ourselves the task to prove that the apparent irregularities of the five planets, the sun and moon can all be represented by means of uniform circular motions, because only such motions are appropriate to their divine nature ... we are entitled to regard the accomplishment of this task as the ultimate aim of mathematical science based on philosophy.

—Claudius Ptolemy, *Almagest I*

Chapter 4

Stellar Aberration

4.1 Appreciation of the phenomenon

It may seem uncultured to interrupt the oleaginous flow of theory by obtruding an aspect of the real world ... but physics is about such interruptions and even physicists cannot indefinitely postpone occasional contacts with physics. I adopt a disrespectful attitude toward the efforts of my fellow laborers in the vineyard of physical description because most have a side of them of which they are unaware—a side in urgent need of a thorn. And the thorn is precisely the subject matter of this chapter, stellar aberration (SA).

Why should a physicist care about SA? (The great majority don't.) First, because it is one of the few examples of genuine one-way propagation of light, so it just might have something to teach us about that interesting subject. More importantly, the facts of SA turn out to be essentially irreconcilable with special relativity theory (SRT), the rock on which the church of modern physics is founded. This is such a darkly guarded secret that you may be reading of it here for the first time. I wonder if I have earned enough credit with you to hope for your attention while I outline the facts and prove the irreconcilability. That will be one of the main goals of this chapter. As a reason for investigating SA more cogent to our present studies, it will be recalled from Chapter 2 that Hertzian (first-order) electromagnetism failed to describe the phenomenon—and our claim that neo-Hertzian (higher-order) theory could do better was only promissory. So, it remains to deliver on the promise.

Let us begin with the basics. Stellar aberration is a phenomenon first observed by James Bradley[4.1] (in 1728) at first order in (v/c). Hence it is generally treated as "classical" in nature. This view is epitomized in Arthur Eddington's explanation of the phenomenon by his famous "umbrella" analogy: a man with an umbrella, running at speed v through raindrops falling vertically at speed c, must tilt his umbrella forward through an angle whose tangent is v/c in order to stay as dry as possible. This deduction is facilitated by consulting a vector triangle whose perpendicular sides are v and c, and whose hypotenuse is therefore of length $\sqrt{c^2 + v^2}$. The result is correct for rain, and the model rightly suggests that *forward* in the sense of the Earth's motion in its orbit is the proper direction to tilt the telescope. Moreover, to first order (given proper interpretation of v), the suggested angle of tilt is correct. However, the fact that the hypotenuse of the vector triangle, representing the relative speed of light traveling down the telescope tube, exceeds c should alert us that something is amiss. In fact, if that leg of the triangle is shortened to match the presumed speed limit c, this closes up the aberration angle to zero. Thus the phenomenon is not explained at all, compatibly with the existence of a limit on the speed of light relative to the telescope tube. So, Eddington's model does not generalize from rain to photons, and is in fact quite misleading in its application to starlight.

Another attempt to concoct a first-order model of the phenomenon was made by Bergmann[4.2] (in 1946) as follows: " ... when a light ray enters the telescope, let us say from straight above, the telescope must be inclined ... so that the lower end will have arrived straight below the former position of the upper end by the time that the light ray has arrived at the lower end." If that were true, then filling the telescope with water, which has a refractive index greater than unity and hence slows the light passing through the tube to speed $c' < c$, should observably change the SA angle from $\tan^{-1}(v/c)$ to $\tan^{-1}(v/c')$. The experiment was done by Airy[4.3] in 1871, with no such result: the water was found to have no effect whatever on SA. It seems that the harder people try to over-simplify the phenomenon by classical models, the less they understand it. (We may add that Berg-mann's book[4.2] had Einstein's approval, and that throughout his life Einstein appears to have over-simplified the SA phenome-non … and this applies to almost all modern physicists as well.)

The fact is that starlight gives us a tantalizing glimpse into the "most quantum" phenomenon in nature, the large-scale non-local action of the quantum of light. We shall never get closer to a personal apperception of the quantum world, and of the meaning of quantum non-locality, as well as of localized *process completion*, than we do through the act of viewing starlight. The physicist, if he is to enter into the quantum world with any degree of under-standing, must bare his head and acknowledge that he stands in the presence of something recognizable (for his professional pur-poses) as holy. In that way he can get over his temptation to trivi-alize the phenomenon by classical models—including SRT (which in its handling and entire conceptualization of light is purely classical). A first step toward grasping the cogency of this last observation is to review what SRT has to say on the subject. We shall find that SRT, in its own mathematically impeccable way, goes every bit as wrong physically as either of the two clas-sical attempts mentioned above.

4.2 SA according to SRT

In this section we refresh the reader's memory regarding well-known textbook material. Our treatment, following Aharoni,[4.4] is ponderous but systematic and rigorous. SRT describes a one-way propagating plane-wave or ray of starlight in an inertial system S by a four-vector $k_\mu = \left(\bar{k}, i\omega/c \right)$, $\mu = 1, \cdots, 4$, where $\bar{k}$ is a three-vector of magnitude $k = 2\pi/\lambda = 2\pi\nu/c = \omega/c$, directed along the wave normal, and $\nu = \omega/2\pi$ is the light frequency (as affected by Doppler effect). In studying SA we may focus attention on mono-chromatic radiation idealized as plane waves, in recognition of the great distances of all stellar sources from Earth. For simplicity we treat our telescope (at rest in S) as situated above the atmos-phere of a uniformly-moving non-rotating Earth, to avoid all con-cerns with non-inertiality, or with refraction and dispersion, con-sequently with distinctions among phase velocity, group velocity, energy velocity, *etc.* (The SRT stipulation of strict inertiality is misleading, because the physics demands non-inertiality of detec-tor motion in order for SA to become a measurable phenomenon.)

As with all four-vectors, the sum of the squares of the four components is an invariant, in this case zero. Thus $\sum_\mu k_\mu k_\mu = k_\mu^2$ $= \vec{k} \cdot \vec{k} + (i\omega/c)^2 = k^2 - \omega^2/c^2 = 0$ is a null vector. This merely expresses the fact that $ck = \omega$ or $\lambda v = c$ in vacuum. Propagation in vacuum takes the photon from its source atom in the star to its earthly detector, which we suppose to be another atom in the photodetector of a telescope sharing the state of motion of the Earth. Our given wave-normal or ray of starlight is described by $\left(\vec{k}, i\omega/c\right)$ in an inertial system S and by $\left(\vec{k}', i\omega'/c\right)$ in another inertial system S'. For the moment we refrain from identifying these systems physically, but shall merely turn the crank of the formalism. According to SRT, the two four-vectors have components that can be related by the special Lorentz transformation or boost,

$$k_1' = \gamma\left[k_1 + i(v/c)k_4\right]$$
$$k_2' = k_2$$
$$k_3' = k_3 \tag{4.1}$$
$$k_4' = \gamma\left[k_4 - i(v/c)k_1\right],$$

where $\vec{k} = k_1, k_2, k_3$, $k_4 = i\omega/c$, $\gamma = 1/\sqrt{1 - v^2/c^2}$, etc., v is the velocity of S' relative to S, and the relative motion of S and S' is directed along their shared x-axes. Invariance of the null four-vector is preserved in S', $\vec{k}' \cdot \vec{k}' + (i\omega'/c)^2 = k'^2 - \omega'^2/c^2 = 0$.

We follow SRT textbook procedures[4.4] to learn how the k_μ four-vector transforms: It is convenient to introduce direction cosines $\ell = \cos\alpha$, $m = \cos\beta$, $n = \cos\gamma$ in S, so that $\vec{k} = k_1, k_2, k_3$ $= k\ell, km, kn$, where $\ell^2 + m^2 + n^2 = 1$. Since $k = \omega/c$, we can write our four-vector in S as $ck_\mu = \omega\ell, \omega m, \omega n, i\omega$. In S' the same light ray is described by $ck_\mu' = \omega'\ell', \omega'm', \omega'n', i\omega'$. The Lorentz transformation (4.1) then yields

$$\omega'\ell' = \gamma\left[\omega\ell + (iv/c)(i\omega)\right] = \gamma\omega(\ell - v/c) \tag{4.2a}$$
$$\omega'm' = \omega m \tag{4.2b}$$
$$\omega'n' = \omega n \tag{4.2c}$$
$$\omega' = \gamma\omega(1 - \ell v/c). \tag{4.2d}$$

Eq. (4.2d) expresses the SRT relativistic Doppler effect, which was discussed in Chapter 3, and provides the derivation of Eq. (3.58) used there. From Eqs. (4.2) we obtain the transformation formulas for the direction cosines,

$$\ell' = \gamma \frac{\omega}{\omega'}(\ell - v/c) = \frac{\ell - v/c}{1 - \ell v/c} \qquad (4.3a)$$

$$m' = \frac{\omega}{\omega'} m = \frac{m}{\gamma(1 - \ell v/c)} \qquad (4.3b)$$

$$n' = \frac{\omega}{\omega'} n = \frac{n}{\gamma(1 - \ell v/c)}. \qquad (4.3c)$$

It is a simple algebraic exercise to verify that $\ell'^2 + m'^2 + n'^2 = 1$, so the transformed quantities are indeed direction cosines in S'. Also, we note that when multiplied by c these relations reduce to the light velocity composition laws obtained by differentiating the Lorentz transformation equations (Møller[4.5]). From these general relations we can determine at once how much the light ray or telescope axis changes its direction in 3-space in the change from S' to S. By vector analysis, or from the geometry of direction cosines, we know that the cosine of the angle α between the two unit vectors $\vec{k}/k$ and $\vec{k}'/k'$ (i.e., between the wave normals in S and S') is the scalar product $\ell\ell' + mm' + nn'$. This "classical" relation holds because under the special Lorentz transformation the spatial coordinate axes in S and S' retain their parallelism, so directions and changes of direction in the two spaces are unambiguously defined by projections onto the coordinate axes. Using $m^2 + n^2 = 1 - \ell^2$, we find from (4.3)

$$\alpha_{SRT} = \cos^{-1}\left[\frac{\ell(\ell - v/c)}{1 - \ell v/c} + \frac{m^2}{\gamma(1 - \ell v/c)} + \frac{n^2}{\gamma(1 - \ell v/c)} \right]$$

$$= \cos^{-1}\left[1 - \frac{1 - \ell^2}{1 - \ell v/c}\left(1 - \sqrt{1 - v^2/c^2}\right) \right] \approx \qquad (4.4)$$

$$\sqrt{1 - \ell^2}\left(\frac{v}{c}\right) + \frac{\ell\sqrt{1 - \ell^2}}{2}\left(\frac{v}{c}\right)^2 + \frac{\sqrt{1 - \ell^2}(1 + 2\ell^2)}{6}\left(\frac{v}{c}\right)^3 + \cdots$$

The expansion of arc cosine in powers of v/c here is a bit tricky and rather interesting, but since the interest is purely mathematical we shall take the result as given.

For historical interest, it may be worth a brief digression to compare with Einstein's original result. In his 1905 paper[4.6] he gave the formula, essentially a velocity composition law,

$$\cos\phi' = \frac{\cos\phi - \beta}{1 - \beta\cos\phi}, \quad \beta = \frac{v}{c},$$

which is very elegant but not very useful. We need to know the telescope tilt angle $\alpha_{SRT} = \phi' - \phi$. The following strategy is as good as any for worming out this information. Let $q = \cos\phi$, $q' = \cos\phi'$. A Pythagorean right triangle with small vertex angle ϕ' can be drawn, with hypotenuse $1 - \beta q$ and long side $q - \beta$. The third (short) side is then of length $\sqrt{1-\beta^2}\sqrt{1-q^2}$, so

$$\tan\phi' = \frac{\sqrt{1-\beta^2}\sqrt{1-q^2}}{q-\beta}.$$

Einstein's formula shows the symmetry $\phi' \leftrightarrow \phi$, $q' \leftrightarrow q$, $\beta \leftrightarrow -\beta$, so

$$\tan\phi = \frac{\sqrt{1-\beta^2}\sqrt{1-q'^2}}{q'+\beta}.$$

Then a standard trigonometric identity[4.7] states that

$$\tan\phi' - \tan\phi = \frac{\sin(\phi'-\phi)}{\cos\phi'\cos\phi},$$

whence

$$\sin(\phi'-\phi) = \sin\alpha_{SRT} = qq'\left(\frac{\sqrt{1-\beta^2}\sqrt{1-q^2}}{q-\beta} - \frac{\sqrt{1-\beta^2}\sqrt{1-q'^2}}{q'+\beta}\right).$$

On eliminating q' by means of Einstein's composition relation above, viz.,

$$q' = \frac{q-\beta}{1-\beta q},$$

we get an expression for α_{SRT} as an arc sine equivalent to the arc cosine expression (4.4). Both show that $\alpha_{SRT} = \phi' - \phi$ can be represented as the small vertex angle of a Pythagorean right triangle with hypotenuse $1 - \beta q$, long side

$$1 - \beta q - (1 - q^2)(1 - \sqrt{1-\beta^2}),$$

and short side

$$\sqrt{1-q^2}\left[\beta - q(1 - \sqrt{1-\beta^2})\right].$$

Einstein's formulas are consistent with these results when q is replaced by his $\cos\phi$. When q is replaced by $\ell = -\sin\theta\cos\phi$ they

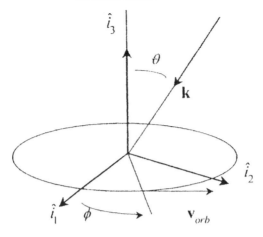

Fig. 4.1. Earth's orbit in plane $\hat{i}_1, \hat{i}_2$ of the ecliptic, showing Earth's position at azimuth ϕ and orbital velocity $\vec{v}_{orb}$; also fixed $\vec{k}$-vector of starlight propagation at polar angle θ, lying in fixed plane $\hat{i}_2, \hat{i}_3$ normal to the ecliptic.

are consistent with angles more generally defined as in Fig. 4.1, so that

$$\ell, m, n = -\sin\theta\cos\phi, \sin\theta\sin\phi, -\cos\theta, \qquad (4.5)$$

the signs signifying downward propagation of the starlight. Opposite signs could equally well be used, signifying upward pointing of the telescope. Although only one parameter, $\ell = -\sin\theta\cos\phi$, appears in the expansion (4.4), it depends on two angles, the polar angle θ and the azimuthal or orbital angle ϕ, the latter varying with the time of year.

It is well known (since Bradley) that over the course of a year any given star traces out on the celestial sphere a small ellipse of semi-major axis a and semi-minor axis b, known as the "figure of aberration." This ellipse is roughly the projection along the mean direction to the star of the Earth's (approximately circular) orbit onto the celestial sphere. Thus for a star near the zenith (θ small) the figure of aberration is almost a circle, whereas toward the horizon (θ near $90°$) it becomes flattened into a line. Holding θ constant at some intermediate value, we note that as the Earth progresses counter-clockwise around its orbit (ϕ varying) the star's position on the projected elliptical figure of aberration advances synchronously, but with a $90°$ phase advance—i.e., in phase quadrature. When $\phi = 0°$ we see from Fig. 4.1 that the Earth is advancing directly toward the star, and that the aberration angle α is a minimum, corresponding to the b-value (semi-minor axis) of the projected ellipse, $\alpha_{min} = b$. Three months later, when $\phi = 90°$, α is a maximum, corresponding to the semi-major axis $\alpha_{max} = a$. These extremal properties can be verified directly from the coefficient $\sqrt{1-\ell^2}$ of the dominant term in (4.4): For $\phi = 0°$, $\sqrt{1-\ell^2} = \sqrt{1-(-\sin\theta)^2} = \cos\theta$, which is minimal for the given θ-value; and when $\phi = 90°$ we have $\ell = 0$ and $\sqrt{1-\ell^2} = 1$, which is maximal.

A convenient gauge of stellar aberration that should be readily measurable is the ratio of semi-axes, as predicted by (4.4); $viz.$,

$$\frac{b}{a} = \frac{\alpha_{min}}{\alpha_{max}} = \frac{\alpha(\ell = -\sin\theta)}{\alpha(\ell = 0)}$$

$$\approx \cos\theta - \frac{\cos\theta\sin\theta}{2}\left(\frac{v}{c}\right) + \frac{\cos\theta\sin^2\theta}{3}\left(\frac{v}{c}\right)^2$$

$$- \frac{\cos\theta\sin\theta\left(\frac{1}{6}+\sin^2\theta\right)}{4}\left(\frac{v}{c}\right)^3 \qquad (4.6)$$

$$+ \frac{\cos\theta\sin^2\theta\left(\frac{2}{9}+\sin^2\theta\right)}{5}\left(\frac{v}{c}\right)^4 + \cdots$$

The leading term here, $\cos\theta$, corresponds to the projection of the Earth's "circular" orbit on the celestial sphere. Owing to the ellipticity of the actual orbit, these calculations would have to be modified in application (and small diurnal variations allowed for), but the circular-orbit approximation, used throughout this chapter, will illustrate the features we wish to emphasize.

In particular, the crucial point is that SRT predicts a first-order departure of the *shape* of the figure of aberration from the simple classical Earth's-orbit projection recipe $(\cos\theta)$, corresponding to Bradley aberration. This departure term has a maximum magnitude at $\theta = 45°$ above the plane of the ecliptic. Since the angular figure of aberration itself is of the order (v/c) radian [*cf.* Eq. (4.4)], to verify alteration of its shape requires measurements of order $(v/c)^2$. Writing in 1946, Bergmann[4.2] stated that "the relativistic second-order effect is far below the attainable accuracy of observation." But since then the observational situation has changed dramatically. With the advent of the Very Long Baseline Interferometry (VLBI) system, it is claimed that angles can be resolved to around 10^{-4} arc sec, or roughly 5×10^{-10} radian. The value of (v/c) is about 10^{-4} radian for the earth's orbital motion. Therefore $(v/c)^2$ effects, of order 10^{-8} radian, should be measurable.

An opportunity thus arises for astronomers to perform a crucial test of SRT in an area (special relativistic shape modification of the classical Bradley figure of aberration) not hitherto probed, so far as this writer knows. To do this would be an important contribution to physics. However, some form of the VLBI system has been in existence since the late 1960's, and nobody has suggested using it to check the second-order term in Einstein's prediction, Eq. (4.4). Why should they? Einstein was right about everything else, how could he be wrong about that? If you offered ten randomly-selected physics professors the chance to use the VLBI system for such a purpose, none would be interested—no career advancement opportunities there. Why face the possibility of failure to obtain the politically correct answer—when failure means instant pariah-hood? Besides, they know the answer from a theory they trust completely. What interests them is the unknown, the frontier. It is not unlike the case of the savants who declined to look through Galileo's telescope. I have not checked, but feel sure they were academicians to a man. We note an ageless pattern: Known theory trumps empirical inquiry every time.

The above derivation is conducted in typical textbook style, in that it leaves the impression that whatever a bunch of mathematics has to say about the phenomenon is all that needs to be said. In this case that is far from the truth. Much more needs to be said, as we shall soon see.

4.3 SA according to neo-Hertzian theory

To review the preceding chapters in one sentence: Starting from Maxwell's field equations as a covered theory, successive improvements in the scope of "invariance" can be achieved by the formal replacements

$$\frac{\partial}{\partial t} \quad \rightarrow \quad \frac{d}{dt} = \frac{\partial}{\partial t} + \vec{v}_d \cdot \vec{\nabla} \quad \rightarrow \quad \frac{d}{d\tau_d} = \gamma_d \frac{d}{dt} = \gamma_d \left(\frac{\partial}{\partial t} + \vec{v}_d \cdot \vec{\nabla} \right),$$

the arrows representing progression from Maxwellian to Hertzian to neo-Hertzian formulations of field theory. In the discussion that follows we shall be concerned with neo-Hertzian predictions alone. The central result from Chapter 3, Eq. (3.49),

was that the neo-Hertzian wave equation has a d'Alembertian solution for which

$$u = \pm\sqrt{c^2 - v_d^2} + \left(\vec{k}/k\right) \cdot \vec{v}_d, \tag{4.7}$$

where u is light (*in vacuo*) propagation phase velocity and $\vec{v}_d$ is
detector velocity with respect to the observer's inertial frame.
Maxwell's result, $u = \pm c$, corresponds to the particular case in
which the light detector is at rest in the observer's frame ($\vec{v}_d = 0$).
To get a feeling for the situation, let us first apply (4.7) to the special case of aberration of light from a star at the zenith. The present discussion is based on previous work by the author.[4.8]

First off, we have the problem, in defining $\vec{v}_d$, of where to
place the "observer" in order to specify "the observer's inertial
frame." For most purposes of ordinary observation the earthly
lab or observatory suffices as sufficiently "inertial;" but in the
case of stellar aberration, as we have already noted, it happens
that the phenomenon achieves observability only through the
slow annual changes resulting from non-inertiality of Earth motion. So we must place our observer in a reference system more
nearly inertial than that of the orbiting Earth. Referring things,
then, to the more nearly inertial system of the Sun (or the solar
system barycenter), we have

$$\vec{v}_d = \vec{v}_{orb}, \tag{4.8}$$

where $\vec{v}_d$ is the velocity of our detector (telescope) at rest on the
Earth's surface and $\vec{v}_{orb}$ is the tangential velocity in the Sun's inertial system of the Earth (neglecting diurnal effects) in its orbit,
approximated as circular. This orbit defines the plane of the ecliptic, to which our ray of starlight $\vec{k}$ from the zenith is normal.
Thus $\vec{k} \cdot \vec{v}_d = 0$ in (4.7), with choice of the plus root, yields

$$u = \sqrt{c^2 - v_d^2}. \tag{4.9}$$

We see that the light, propagating straight down from the zenith,
is slightly slowed by a second-order amount. We know also that
for an observer at rest with respect to the telescope tube $\vec{v}_d$ vanishes and $u = c$, in agreement with Maxwell. What can this mean?
Only one possible thing: for the light to have speed c relative to
the telescope tube and speed $\sqrt{c^2 - v_d^2}$ relative to the vertical (zenith) direction, *the telescope tube has to be tilted away from the vertical.* This tilting is evidence of "stellar aberration." It is a phenomenon whose existence is clearly predicted *ab initio* by the line

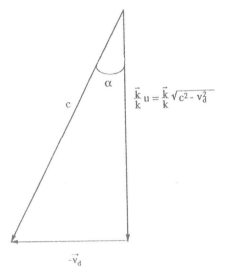

Fig. 4.2. Vector diagram (Pythagorean triangle) showing tele-scope tilt angle α for starlight from the zenith, as described by neo-Hertzian electromagnetism.

of reasoning given here. The predicted angle of tilt for starlight from the zenith, as shown by the vector triangle of Fig. 4.2, is

$$\alpha_{neo-Hz} = \sin^{-1}\left(v_d/c\right) = \tan^{-1}\left(v_d/\sqrt{c^2 - v_d^2}\right) \approx \frac{v_d}{c} + \frac{1}{6}\left(\frac{v_d}{c}\right)^3 + \cdots, \quad (4.10)$$

and we may take the sense of tilt to be "forward" in the direction of Earth's orbital velocity, to agree with Bradley's observations. (The present argument fails to specify the sense of the telescope tilt, which can nevertheless be deduced from elementary considerations.) Since the wave-normal $\vec{k}$-vector is vertical [and is not affected by inertial transformations, as the invariance $\vec{k}' = \vec{k}$ of Eq. (3.54) shows], filling the telescope tube with water does not alter the horizontal orientation of the wave-fronts. It does slow the propagation along the inclined tube direction in the same ratio as the vertical descent, so the Pythagorean diagram, Fig 4.2, is not changed, and the tilt angle is unaffected, in agreement with the observations of Airy.[4.3] We recall that the classical diagram corresponding to Fig. 4.2, based on the Eddington "umbrella" model mentioned above, is also Pythagorean, but with a vertical side of length c and an hypotenuse of (impossible) length $\sqrt{c^2 + v_d^2}$, so that

$$\alpha_{classical} = \tan^{-1}\left(v_d/c\right) \approx \frac{v_d}{c} - \frac{1}{3}\left(\frac{v_d}{c}\right)^3 + \cdots \qquad (4.11)$$

All theories agree as to the first-order term. Note that neo-Hertzian theory, according to Eq. (4.9) or (4.10), specifically identifies *detector* velocity as the parameter appropriate to the description of SA. Neither light source speed nor source-sink relative speed enters. Some other theories, such as SRT, are less clear about identifying what "relative velocity" is involved. We shall say more about this presently. By coincidence it happens for our special case of starlight from the zenith that SRT agrees to third order with the neo-Hertzian result (4.10)—as follows from Eq. (4.4) with $\theta = 0$, $\ell = 0$. However, such agreement does not hold in general.

Our use of the Sun as a referent in the above discussion was merely for simplicity of presentation. More generally, we may consider an initial (pseudo-) inertial system S that co-moves with the telescope at time t_0 and a second similar system S' that co-moves with it at later time t_1. Then if $\vec{v}_d$ is interpreted as the velocity of S' relative to S, the formula (4.10) and analysis remain valid with this new meaning of the symbols. The "tilt angle" so described is the angle of *tilt change* between changes of inertial system. Thus we are freed from concern about the Sun and can recognize that *only detector motions need be considered* in giving a complete empirical description of the SA phenomenon. Such a Sun-free description has been recommended by Synge.[4.9]

A particular virtue of this perception is that it reemphasizes and forces us to recognize that the observability of SA arises purely through *non-inertiality* of the detector's motion. If our telescope remained at rest permanently in an inertial system, or if any integral multiple of 12 months marked the time interval $t_1 - t_0$, SA would be unobservable. Thus, if S and S' are the same inertial system, $\vec{v}_d = 0$ and from (4.10) $\alpha_{neo-Hz} = 0$; so there is no observable aberration (*i.e.*, no *change* of aberration angle). One could write $\Delta\alpha$ instead of α, to be more explicit about this. [Here, though, we may view α alternatively as the "constant of aberration," or angular radius of the near-circular figure of aberration traced out on the celestial sphere over a year's time by the image of a star at the zenith. For this purpose Eq. (4.8) applies.]

Let us now work out the neo-Hertzian analysis for the general case of a star not at the zenith. From the geometry of Fig. 4.1 we see that

$$\vec{k}/k = -\hat{i}_2 \sin\theta - \hat{i}_3 \cos\theta \qquad (4.12)$$

and

$$\vec{v}_d = -\hat{i}_1 v_d \sin\phi + \hat{i}_2 v_d \cos\phi . \qquad (4.13)$$

Hence

$$\left(\vec{k}/k\right)\cdot\vec{v}_d = -v_d \sin\theta\cos\phi , \qquad (4.14)$$

so that Eq. (4.7) yields

$$u = \sqrt{c^2 - v_d^2} - v_d \sin\theta\cos\phi . \qquad (4.15)$$

We may consider the telescope axis to be pointed along the direction of a "telescope vector" $\vec{T}$, which is the negative of the vector sum $\left(\vec{k}/k\right)u$ and $-\vec{v}_d$, just as in the velocity composition diagram of Fig. 4.2. Then

$$\vec{T} = -\left[\left(\vec{k}/k\right)u - \vec{v}_d\right]$$
$$= u\left(\hat{i}_2 \sin\theta + \hat{i}_3 \cos\theta\right) - \hat{i}_1 v_d \sin\phi + \hat{i}_2 v_d \cos\phi \qquad (4.16)$$
$$= -\hat{i}_1 v_d \sin\phi + \hat{i}_2 \left(u\sin\theta + v_d \cos\phi\right) + \hat{i}_3 u\cos\theta$$

where u is given by (4.15). The magnitude of $\vec{T}$ is

$$T = \sqrt{\vec{T}\cdot\vec{T}} = \sqrt{u^2 - 2u\left(\vec{k}/k\right)\cdot\vec{v}_d + v_d^2}$$
$$= \sqrt{c^2 - v_d^2 \sin^2\theta\cos^2\phi}. \qquad (4.17)$$

The scalar product of unit vectors $\vec{T}/T$ and $\vec{k}/k$ is the cosine of the angle between these vectors, which is our telescope tilt angle (or rather, tilt angle change associated with change of velocity vector $-\vec{v}_d$, as discussed above),

$$\alpha_{neo-Hz} = \cos^{-1}\left[\left(\vec{T}\cdot\vec{k}\right)/Tk\right]. \qquad (4.18)$$

Since, with the help of (4.15), we have

$$\vec{T}\cdot\left(\vec{k}/k\right) = \left[-\hat{i}_1 v_d \sin\phi + \hat{i}_2 \left(u\sin\theta + v_d \cos\phi\right) + \hat{i}_3 u\cos\theta\right]$$
$$\cdot\left[-\hat{i}_2 \sin\theta - \hat{i}_3 \cos\theta\right] = -u - v_d \sin\theta\cos\phi = -\sqrt{c^2 - v_d^2}. \qquad (4.19)$$

Eq. (4.18) with (4.17) yields [exploiting the sign ambiguity of the radical in (4.7)]

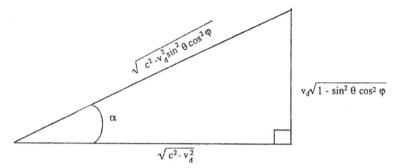

Fig. 4.3. Pythagorean triangle for telescope tilt angle α in the neo-Hertzian general case of starlight propagating with direction cosine $\ell = -\sin\theta\cos\phi$.

$$\alpha_{neo-Hz} = \cos^{-1}\sqrt{\frac{c^2 - v_d^2}{c^2 - v_d^2\sin^2\theta\cos^2\phi}}.\qquad(4.20)$$

For purposes of easiest expansion, this is advantageously re-expressed as

$$\alpha_{neo-Hz} = \tan^{-1}\left(\frac{v_d}{c}\sqrt{\frac{1-\ell^2}{1-(v_d/c)^2}}\right)$$

$$\approx \sqrt{1-\ell^2}\left(\frac{v_d}{c}\right) + \frac{\sqrt{1-\ell^2}\left(1+2\ell^2\right)}{6}\left(\frac{v_d}{c}\right)^3\qquad(4.21)$$

$$+\frac{\sqrt{1-\ell^2}\left(3+4\ell^2+8\ell^4\right)}{40}\left(\frac{v_d}{c}\right)^5+\cdots$$

where $\ell = -\sin\theta\cos\phi$, as in Eq. (4.5), and the angles are those of Fig. 4.1. The simple trigonometry governing passage from (4.20) to (4.21) is that of the Pythagorean right triangle shown in Fig. 4.3. Study of (4.21) reveals the same phase effect as was noted for SRT, such that the motion of the stellar image along the annual figure of aberration on the celestial sphere is in phase quadrature with the Earth's position in its orbit (leading by $90°$). Further discussion is given in earlier work by the author.[4.8]

We record the neo-Hertzian result analogous to the SRT prediction (4.6), *viz.*,

$$\frac{b}{a} = \frac{\alpha_{\min}}{\alpha_{\max}} = \frac{\alpha(\ell = \sin\theta)}{\alpha(\ell = 0)} = \frac{\tan^{-1}\left(\dfrac{v_d \cos\theta}{\sqrt{c^2 - v_d^2}}\right)}{\tan^{-1}\left(\dfrac{v_d}{\sqrt{c^2 - v_d^2}}\right)} \tag{4.22}$$

$$\approx \cos\theta + \frac{\cos\theta \sin^2\theta}{3}\left(\frac{v_d}{c}\right)^2 + \frac{\cos\theta \sin^2\theta\left(\dfrac{2}{9} + \sin^2\theta\right)}{5}\left(\frac{v_d}{c}\right)^4 + \cdots$$

The absence of a second-order term in the neo-Hertzian result, (4.21), and the presence of such in the SRT result, Eq. (4.4) [or the absence of a first-order term in (v_d/c) in (4.22) and its presence in (4.6)], is especially to be noted. It provides the basis for a crucial test to decide between the two theories. We pointed out above that, following development of the VLBI system in the late 1960's, it appears to have become technically feasible to make second-order astrometric observations of sufficient refinement to do this. All that has been lacking is the will. No new apparatus is needed, just observation time and some peanut bars to defray the labor of graduate students.

Finally, it may be wondered why Potier's principle, which denied the observability of any distinction between SRT and Hertzian theory in the case of the Sagnac effect, does not do the same for SA. I believe the reason is that Potier's principle refers to a single light path with fixed endpoints; whereas here we may consider that we have either two light paths (at different times) or a single light path with an accelerated endpoint. It is worth mentioning that both Potier's principle and Fermat's principle from which it derives (which states that physical light takes the "least time" in going between two points) are cast into limbo by the advent of SRT. The latter introduces two types of "time," proper and frame. Fermat's principle certainly cannot refer to an interval of proper time, since that is zero on all light paths. But if it refers to frame time, SRT informs us that frame time is "meaningless," nonexistent because non-invariant. Only some *Raffiniert* thing of beauty called "spacetime" is meaningful. So, Fermat's principle seemingly dies without burial—another victim of the great destroyer. In Chapter 6 we shall introduce a type of "collective time" that would allow its reincarnation. I admit I have not given this subject area much thought. The reader may find it worthwhile to do better.

4.4 SRT's unrecognized conceptual difficulties with SA

From Eq. (4.21) it will be seen that the predicted neo-Hertzian constant of aberration depends on v_d, which is detector (telescope) speed and is thus unambiguously linked *via* Eq. (4.8) to Earth's speed in orbit,

$$v = v_d = v_{orb}. \tag{4.23}$$

Consequently it is in agreement with observation. But the corresponding parameter in the SRT prediction (4.4), denoted "v," is more problematic as to its physical interpretation. Einstein's attempt in his 1905 paper[4.6] was disastrous. He simply made a mistake. It was a very natural mistake (and one often repeated by later authorities such as C. Møller[4.5] and numerous others cited in a recent article,[4.10] a revealing study of the confusions about stellar aberration fostered by SRT but endemic in both physics and astronomy). It has far-reaching consequences that even to this day are unrecognized by most scientists. Einstein assumed v to be the relative velocity between one inertial system in which the stellar light source was at rest and another inertial system in which the light detector was at rest. In other words, $\vec{v}$ was *source-sink relative velocity* (radial component, though I shall generally omit this qualifier to save words), of magnitude

$$v = v_{source-sink}. \tag{4.24}$$

This was actually a forced move on his part, (a) because of his general philosophy of "relativism," (b) because the source and sink represent the two *termini* most naturally associated with a light ray, and (c) because he wanted to assert spacetime symmetry properties for light-propagation descriptors; *i.e.*, to treat $\left(\vec{k}, i\omega/c\right)$ in a way formally analogous to $(\vec{r}, ict)$. Now, since the frequency part ω/c of this starlight descriptor is governed by the Doppler effect, which is known to be dependent on source-sink relative speed (as noted in Chapter 3, Section 5), this meant that the same velocity parameter v, with the same interpretation, had to be used in describing the space part $\vec{k}$. The latter described SA—so it appeared to be a "no-brainer" to assume that the v in Eq. (4.4) meant source-sink relative velocity, as in (4.24).

But of course that wasn't empirically correct. The stellar light sources we see are known to be in all sorts of states of motion, implying all sorts of values of source-sink relative velocity $v_{source-sink}$, hence leading necessarily to all sorts of calculated α-

values; yet in fact all stars show the same SA, the same α-value. The great distance of the stars makes no difference because relative velocity is unaffected by distance. The "fixed stars" are a fiction of no interest to physics. There is thus no option to use (4.24) as the velocity parameter describing SA; one is forced by facts to use (4.23), $v = v_d = v_{orb}$. Here is a plain qualitative conflict of SRT ideology with observation. The facts are there and the theory doesn't fit them.

Einstein thus historically made a false prediction of the angle of stellar aberration—but because he got the right first-order *formula*, and modern physicists tend to look only at formulas, few noticed that his *interpretation* led to a false use of the formula. Did Einstein ever publish a correction? Not that I know of. Still, one infers from indirect evidence that he did at some later time tumble to a recognition that his tune had to be changed in regard to SA. Thus in the 1946 book by Bergmann,[4.2] which had Einstein's approval and to which he contributed a Foreword, the Sun has quietly crept in as a replacement for the light source—so that now v represents Sun-sink, rather than source-sink, relative velocity. Sneaky, eh? We still use "relative" velocities, but in this shell game have smoothly switched one of the referents to an innocent bystander, the Sun. Quietly, quietly ... so as not to frighten the horses. A mere prick, you say, yet observe its effect: not so deep as a well, nor so wide as a church door; but 'tis enough, 'twill serve. It reaches to the core and draws the heart's blood of spacetime symmetry. For if spacetime symmetry holds, it must be that the quiet switch of referents on the spacelike side applies symmetrically and equally quietly on the timelike side, the frequency or Doppler side.

But that puppy will neither woof nor whine. It is too quiet to live. For obviously nobody has described the Doppler effect *without* using what amounts to source-sink relative velocity—more exactly, the radial component thereof. The Sun-sink system is totally irrelevant. So, where do we stand? (1) It is essential for timelike Doppler frequency description—of the sort on which rests the hypothesis of the "expansion of the universe"—to use a numerically *different* velocity parameter ($v_{source-sink}$) from the one that is essential to use for spacelike SA description (v_{orb}). (2) It is essential for the validity of spacetime symmetry and universal covariance that *the same* velocity parameter be used in transforming *both* spacelike and timelike parts of any SRT four-vector—because a four-vector by definition obeys the Lorentz transformation, and the Lorentz group of such transformations is a one-parameter group, that parameter being the relative velocity of two and only two inertial systems. (3) It is essential to the ideology of SRT that starlight be described by a four-vector. (4) It is impossible that SRT be wrong about anything, because we have been repeatedly informed by experts of SRT's apotheosis: is no longer a theory but a fact. It is the fact of facts. Having been created by a demi-god, it is more factual than facts, more factual than the universe. When the universe has faded to mere shadows, SRT will still be there, facting away. No wonder everybody keeps quiet. It's the only solution for slick tenure and a well-greased career: Universal covariance supported by universal quietness. And it works. It has worked worldwide for a century ... and no reason not to work for ten more: Ptolemy's folly worked that long. SRT has acquired the necessary critical mass of quietness behind it to last 'til the cows come home.

4.5 Einstein's state of mind: a speculation

This is written in the centenary year, 2005, of Einstein's *annus mirabilis*, and he is much in the public mind, and in mine as well. I cannot resist speculating that his final years must have been secretly unhappy ones. From the very silence he maintained about SA over all those years, one is free to imagine that he hoped nobody would bring up the subject. And apparently nobody ever did. Any strategy is good that works. He must have known, certainly no later than 1946 when the overt switch from $v_{source-sink}$ to v_{orb} occurred, that the jig was up for spacetime symmetry. So, I will hazard the guess that in his later years he knew the whole subject of SA was sudden death for SRT ... for he had long been quoted as saying that a single observational failing would doom his or any other physical theory. And what was this but the smoking gun? — an obvious stress failure of the linchpin of all his theorizing, universal covariance. At symposia he must have lived in fear of the wrong question from some brash graduate student in his audience. But he lucked out to the end ... nobody was the least bit curious about SA. Nobody ever asked

> *The question:* How can *both* SA and Doppler effect be part of *the same* four-vector of starlight?

And that is true to this day. Einstein may have been unhappy about the durability *sub specie aeternitatis* of his system, as one of his last letters to a friend suggests, but he was also lucky. To keep the world at one's feet for so many years, both before and after one's death, it helps to be not only wise but lucky. Alternatively, maybe he never read the elementary parts of Bergmann's book, but just dashed off a Foreword and lived out his life in blissful ignorance of the booby trap buried in the foundational bedrock of his world system. Let us hope so ... for in a world glutted with information, groaning under the unwanted datum that George Washington had teeth of wood, it would be a telling blow to learn that Einstein had feet of clay.

4.6 A rebuttal

When all has been said, there remains the inevitability of a rebuttal from the SRT camp. Relativists have repeatedly shown themselves skillful navigators to any port in a storm. I find it easiest to exemplify this virtuosity in terms of an imagined dialogue between myself and a hypothetical physics professor:

> PHYSICS PROFESSOR: I think in your eagerness to make rhetorical points you have overlooked a simple fallacy in the argument you used to attack SRT, and in particular to "disprove" spacetime symmetry.
>
> ME: What's that?
>
> PP: Do you recall your claim that "It is absolutely essential for timelike Doppler frequency description to use a numerically *different* velocity parameter ($v_{source-sink}$) from the one that is absolutely essential to use for spacelike SA description (v_{orb})."?
>
> ME: Indeed.
>
> PP: Permit me to point out that I am at perfect liberty to use v_{orb} in describing the Doppler effect. I would not normally do this. Nobody in his right mind would. But when the existence of spacetime symmetry is at stake, on which all of advanced modern physics securely rests, extreme measures are justified.
>
> ME: What is the outcome of that parameter choice?
>
> PP: The resulting "Doppler effect" shows a small annual variation affecting all spectral lines, due to changes of Earth's velocity relative to each stellar light source. All information that differentiates stellar sources—all informa-

tion about spectroscopic differences between near and far stars, *etc.*—is lost, because we have given up access to information on the identities of the various spectral lines. But *all lines* undergo annually a small identical frequency shift that tracks the Earth's orbitings.

ME: So spectral frequencies no longer identify the chemical elements, and the "Doppler shift" no longer identifies radial velocities of stellar sources, but, *symmetrically with SA*, the latter shows a small annual cyclic change, due to telescope motion, that is the same for all stars?

PP: Exactly. This proves spacetime symmetry.

ME: Hmmm … I see your point. Frankly, I had not expected the Expanding Universe to be sacrificed on the altar of SRT's ideology, but now that you have volunteered such a trade-off I can see its efficacy. I assume you will now want to redo cosmology from the ground up.

PP: No, of course not. I discarded the spectroscopic information only temporarily, to make a point of principle. Naturally, it would be foolish to throw away the data we have from laboratory measurements of spectra of the elements, concerning the unshifted frequencies of spectral lines at zero source velocity, from which we can infer the absolute velocity of stellar source recession, different for each star.

ME: I see. You are saying that a relativist can use one parameter, v_{orb}, when one of his cloud-castles of thought, SRT, is threatened, and another parameter, $v_{source-sink}$, when another of those cloud-castles, the Expanding Universe, is threatened. I conceive of such a relativist less as a scientist than as a chameleon … a Master of Arts of expediency. I perceive you as incorrigibly determined to seize any line of argument to anaesthetize your conscience and enable you to think about other things.

PP: I perceive you as incorrigibly determined to misunderstand science. It was a waste of my time to try to correct your obtuseness, which I seriously believe to arise from some form of dementia.

ME: Fortunately, we do not have to bandy epithets. An objective test of spacetime symmetry, or of covariance *vs.* invariance, is readily available to us. You have only to use your personal clout with your colleagues down the hall in the Department of Astronomy to persuade them to make careful astrometric measurements (using the existing VLBI system) to verify the "Einstein effect," a second-order departure of SA from Bradley aberration predicted by SRT.

This never-verified effect is a beautiful and accurate "signature" of Einstein, written for all time in the indelible stars. It should provide a nice thesis topic. It has only to be looked for. If it is there, I will eat my hat.

PP: Much as I should enjoy that gustatory spectacle, and be happy to supply the mustard or Tabasco, it happens that I have better things to do with my time than to pursue such a foregone conclusion. Everyone knows the inevitable outcome from theory. I prefer not to be laughed at by all who understand modern science.

4.7 Another "first test" failure of $d\sigma$: the rigid body

Einstein may have been troubled further in his later years by another unexpected shortcoming of SRT: not only did it fail its first test against one-way propagating light, it failed its first test against mechanics. Initially, by his own account, he thought his theory *solely* concerned "idealized rigid bodies." Within a few years it turned out that the rigid body (of a type defined by Born as always undergoing the Lorentz contraction in the direction of relative motion) was deficient in physical degrees of freedom, according to a theorem of Herglotz-Noether,[4.11] and thus had to be banished as a physically "impermissible idealization." This discovery must have been for Einstein rather like stepping on a rake. Suddenly, what the theory had solely concerned became impermissible for admission to any physical theory. And, quietly, quietly the big claim became a big bluff: As long as the theory solely concerned rigid bodies (and particles) it could make the big claim to be a covering theory of Newton's mechanics. But, as soon as rigid bodies were driven from the temple, SRT became a covering theory only of Newton's particle mechanics (well ... ahem, almost!), not of his mechanics of extended structures ... not of his first-order physics. But was the big claim to cover the first order ever withdrawn? If so ... quietly.

Advocates still like to assert or let it be inferred that SRT can cover the low-speed regime of mechanics every bit as well as Newton does. But this is not in fact the case. SRT covers classical particle mechanics reasonably well over short distances. [We saw from Eq. (1.5d) that the coverage *fails* over "long" distances $x \sim \left(c^2/v \right) t$, which for t short enough need not be long at all.] This is true because SRT's on-worldline timelike invariant $d\tau$, Eq. (3.1), goes over smoothly in the limit $c \to \infty$ to the local Newtonian first-order descriptor dt. But there is no limit in which the space-like alleged "invariant," $d\sigma = icd\tau$, Eq. (3.6), goes over into anything of conceivable interest to man or beast. Consequently, Newtonian rigid body mechanics is not "covered" at all. It stands naked of higher-order embellishments, or banished to the limbo of engineering. SRT cannot handle rigid bodies even at the lowest speeds, much less at high speeds. This is glossed over by SRT's dedicated apologists through hand-waving about the speed of "sound" in rigid bodies being infinite, which is plainly impossible, so rigid bodies are impossible. Of course rigid bodies are impossible! Newton knew as much. What they are is first approximations to extended structures (with deliberate neglect of *elasticity*). As such they are not only legitimate conceptual objects of scientific study but very useful ones. (Mechanical engineers will surely not be the sole witnesses to testify to this.) As first approximations, descriptive of the limiting case in which elasticity goes to zero, they work perfectly well for many practical purposes at low speeds—and how well they work at high speeds is not a topic of empirical knowledge.

The thing about rigid bodies, and the reason they clash with SRT, is that they express the idea of *spatial extension* in its purest, logically unadulterated form—free of all considerations of cause or mechanism. That is bound not to fit with SRT, because by its basic conception of spacetime symmetry that discipline confounds space with time, thus disallowing pure space any conceptual breathing-room. (Recall the miraculous-magical, insightful-fantastical, mathematical-poetical, guru-revealed *fading to mere shadows* that introduced physicists *en masse* to religious experience.) When you take a pure-spatial object like a rigid body and force it into a spacetime-symmetrical Procrustean bed, something has to give ... and what gives is sanity. For surely it is insane to say, *as physics* (though it is perfectly true as mathematics), that a Born-rigid body "loses degrees of freedom," so that it cannot rotate. Such is merely an over-educated way of saying that we have over-constrained our problem ... and of evading recognition that the over-constraint lies in our imposition of spacetime symmetry upon a physics that will have none of it. Like it or not, space (purely spatial extension) is there as an aspect, an ineluctable part, of nature ... and if we cannot face that part squarely we are going to have to abandon part of our sanity. That is what the "invariance of $d\sigma$ " is about—the surrender of sanity to (a contemporary majority view of) beauty.

A special relativity catechism:

What is a Born-rigid body?

It is any body that undergoes the Lorentz contraction.

What is a metric standard?

It is any body that undergoes the Lorentz contraction.

What idealization is impermissible for admission to any theory?

A Born-rigid body.

So, dear student, what metric standard is admissible to SRT?

To review: SRT begins by analyzing nature in terms of rigorously inertial motions. In dulcet tones of austere logic, it speaks only to inertial motions. Suddenly in treating matters of spatial extension it meets contradictions, inconsistencies, unless it can banish rotations altogether. So, it ends up speaking to rotations, after all, but raucously. The tones are no longer dulcet. And since it cannot banish physical rotations it banishes instead, as "impermissible," the property of rigidity, so that *physical* rotation is allowed to occur only through the lucky intervention of a *physical* property known as elasticity. It's logical, it's talmudic, but is it sane? How far in such a direction will sane people allow themselves to be led? The empirical answer: Oh, quite far.

As a point of methodology, to exclude useful approximations on doctrinaire grounds of forced logic stemming from the need to avoid contradictions within an extraneous pet theory is (in a sane world) to cast serious doubt on the said pet theory. In the present case the doubt is only too well-founded. The classical theory of rigid bodies exists without reference to physical phenomena such as sound, gravity, or light. It makes no pretence to truth. It has no more to do with reality than does the point particle ... both are simply useful approximate conceptions, earning and deserving the respect of physicists for their usefulness, not for their truth. To discard a whole realm of useful classical mechanics because SRT *post hoc* declares its basic approximation "impermissible" is presumptuous folly. SRT would like to claim credit as a covering theory of classical mechanics, and originally did so ... but then discovered a failing in itself that necessitated turning that failing into an asset. This was done by claiming in effect that a non-covering theory was better than a covering theory, because the non-covering theory treated more accurately the behaviour of "sound" or communication among the parts of an extended structure. (Actually, this non-covering theory never amounted to a theory ... it was never more than a theory-spoiler. It did not offer engineers a better replacement for rigid-body mechanics—just a metric construction kit with menaces but no instructions.)

Improved communication among parts? Shucks, classical rigid bodies don't even have parts. Being wholes, of course they misrepresent inter-communications among "parts," which are irrelevant to any holistic conception. Rigid bodies misrepresent all details of solid state physics—and are excellent first-approximate descriptors of extended structures for that very reason. That's what idealizations are for. A policy of attacking useful approximations works against the future of physics by discrediting its past. I have described SRT[2.11] as a forest fire that swept through physics leaving only charred stumps of concepts. To declare a useful idealization impermissible is to place a black mark against the context responsible for the declaration. Would you not say that any theory that claimed to give higher-order descriptive accuracy had *failed its first test* when it failed to "cover" the relevant first-approximate descriptor? In other words such a candidate higher-order descriptor is not even able to meet minimum lower-order descriptive requirements.

Is it plausible that one can whip up out of an imagined "symmetry" an "invariant" descriptor $d\sigma$ of spatially-extended structures that is not even a real measure of space? And on the basis of this useless mental creation banish a useful one? SRT got its foot in the door by mildly promising to give physicists a better description of high-speed motions, while not disturbing the valid descriptions of low-speed motions they already had. It went on grandly to discard those low-speed descriptions it did not like and to stir up in physicists such an ecstatic *frisson* of world-structural insight as to rob them of all traditional caution. And now such revelation has got its foot firmly wedged in the door, where will physics go from here? No need to ask—it has long gone.

Surely (I hear you say) the body of empirical data supporting such a violent departure from all past conservative physical descriptive policy must be utterly crushing ... overwhelming. But no ... look again. The cupboard is bare. There are no data at all. Data have we none. Evidences are there none. For, to get evidence supporting directly any of the claims of SRT regarding extended structures or spacelike description generally, it would be necessary to put an extended structure into motion so near the speed of light as to be technically infeasible. By the slender grace and thread of that infeasibility hangs the whole half of SRT based on invariance of the spacelike interval $d\sigma$. That is the half that tells us all the gee-whiz "facts" about world structure. SRT, we are informed over and over, is the most extensively confirmed of all physical theories. Alas, no! The Emperor lacks not only clothes but legitimate title to the throne.

What exactly is all this endless observational confirmation that earns for SRT the to-the-death loyalty of Achilles' myrmidons and Caesar's legions? With few exceptions it is evidence that supports our timelike invariant $d\tau$, Eq. (3.7). (The few exceptions are negligible as evidence, being based on inference from a contaminated source, Maxwell's equations.) That is, the vast bulk of data supporting SRT are those that confirm the invariance of the proper-time interval, as they show SRT's description of *on-trajectory physics* to be correct. All high-speed single-particle motion description fits this category, all the famous "accelerator design" criteria, even $E = mc^2$. There is nothing in any of this evidence to give a shred of support to any of SRT's statements about $d\sigma$, about extended physical structures, or about world structure. Nor, in fact, is there evidence for the timekeeping *symmetry* (among inertial systems) that is integral to Einstein's theory. On the contrary, as we shall see in later chapters, there is reasonably conclusive evidence against it.

4.8 Newtonian point particle mechanics

Though it appears to be another digression, it is worthwhile to interject here a brief inquiry into the foundations of particle mechanics, in order to reinforce the above point. It requires only three short steps to show that the accurate description of the mechanics of high-speed particles depends only on the invariance of $d\tau$, not of $d\sigma$:

Step 1. Newton's second law specifies the motion of a point particle [correctly at first order in (v/c)] by the formula $\vec{F}_{lab} = m_0 \vec{a}$, or more accurately,

$$\vec{F}_{lab}^{(Newton)} = \frac{d}{dt} m\vec{v} = \frac{d}{dt} m \frac{d}{dt} \vec{r} , \tag{4.25}$$

where $\vec{F}_{lab}^{(Newton)}$ is the traditional Newtonian force measured in the laboratory inertial system.

Step 2. We postulate (*via* $t \to \tau$) a higher-order invariant formal counterpart of (4.25) for single-particle timelike trajectory description, *viz.*,

$$\vec{F}_{inv}^{(timelike)} = \frac{d}{d\tau} m_0 \frac{d}{d\tau} \vec{r} = \gamma \frac{d}{dt} m_0 \gamma \frac{d}{dt} \vec{r} , \tag{4.26}$$

where m_0 is invariant rest mass and use has been made of Eq. (3.5). Note that in the formulation of (4.26) an invariant quantity replaces its non-invariant counterpart in (4.25); that is, $t \to \tau$, $m \to m_0$, in place. For utmost generality, m or m_0 can be considered possibly to vary with laboratory time t in both (4.25) and (4.26).

Step 3. The invariant "force" affecting the trajectory of a particle we define to be related to the force acting in the laboratory by

$$\vec{F}_{inv}^{(timelike)} = \gamma \vec{F}_{lab} . \tag{4.27}$$

(A similar relationship makes its appearance in SRT, where the "Minkowski force"[4.5] plays formally the role of our present "invariant" force.) The reason for appearance of the γ-factor in (4.27) is simply that when invariant proper time enters the analytic expression for force action, as it does in (4.26), a correction is needed for the fact that laboratory measurements of force $\vec{F}_{lab}$ refer to "time" measured by clocks running at a different (frame time) rate. Then from (4.26) and (4.27)

$$\vec{F}_{lab} = \frac{d}{dt} m_0 \gamma \frac{d\vec{r}}{dt} = \frac{d}{dt} m_0 \gamma \vec{v} , \tag{4.28}$$

which allows the "old fashioned" alternative interpretation of mass increase, $m = m_0 \gamma = m_0 / \sqrt{1 - v^2 / c^2}$, if we like; or instead $\vec{v} \to \vec{V} = \vec{v}\gamma$, if we don't like. Physically, mass variation is a perfectly viable and perspicuous concept, despite a recent concerted attack on it by *manifest covariance* fanatics.

It is to be noted that force laws of a "spacelike" character, whose analytic expression makes no reference to time, do not need the clock running-rate correction factor γ and thus transform invariantly [because space variables transform invariantly, per Eq. (3.7)], *i.e.*,

$$\vec{F}_{inv}^{(spacelike)} = \vec{F}_{lab} \,. \tag{4.29}$$

This applies, for example to gravitational forces, which derive from the spatial gradient of a scalar potential function, with no reference to time. We shall find the distinction between (4.27) and (4.29) to be of crucial importance when we come to discuss electromagnetic forces. These two relations are perhaps better viewed as correction rules applicable to all forces (to allow for the general operational impracticality of reading invariant proper time from particle-co-moving clocks) than as "laws of nature."

Eq. (4.28) is what all the so-numerous (and so-redundant) high-speed particle observations confirm. At no point in its derivation was any reference made to $d\sigma$. From all the foregoing we may conclude (a) that Einstein never had an honest covering theory of Newton's mechanics, and (b) that what success he did achieve in mechanics could much more simply have been attained by replacing Newton's non-invariant t with invariant particle proper time, $t \to \tau$, and applying the above three-step manipulation. Such a procedure yields an honest covering theory, because there is no space-side fiddling—and no concomitant paradoxes. In a word, the dread Lorentz contraction is totally exorcised. To open our minds to appreciation of a higher grade of music, we need to lay aside Einstein's fiddle.

4.9 Chapter summary

Having established *via* SA the likely existence of a violation of "spacetime symmetry," we need only blow gently on this whole house of cards to witness its collapse. There never was such a symmetry. It was born of a parametric deficiency of Maxwell's equations, and fails the first test (SA) of ability to describe one-way light propagation (there being no consistent velocity parameterization of SA and Doppler effect that will allow both to coexist as parts of the same four-vector). This alleged symmetry prevents SRT from being a covering theory of classical rigid-body mechanics. It seemed to succeed only through mass psychological persuasive efforts, and a public will to set aside disbelief, analogous to the felt need for religious faith. Operationally, there was never a shred of plausibility in the spacetime-symmetry notion. The instruments and operations needed to perform space and time *mensuration*, show no symmetry whatever. But it is not enough to say that spacetime symmetry does not exist. With reference to Eqs. (4.27) and (4.29), we see that an explicit *asymmetry* of the formal mathematical treatment of space and time needs to be accommodated by physical theory. That the exemplary operationalist and rock of integrity Percy W. Bridgman failed to challenge spacetime symmetry is a wonder of the world. (Instead, he supported it with arguments[4.12] I find wholly unconvincing.) It shows that nobody in its path is immune to the sweep of a religious movement on the march.

We see in sum that the troubles SRT has with SA on the electromagnetic side and with rigid bodies on the mechanical side are complementary aspects of the retribution it earns for postulating a non-physical spacelike "invariant," $d\sigma$. Once the sole intellectual support for this alleged "invariant," Maxwell's theory, is discarded and replaced by a higher-order *invariant* description of electromagnetic phenomena (as in Chapter 3), all justification (indeed, all possibility) for such a non-physical postulate disappears. The mystique of spacetime symmetry—oblivious as it is of all operational warnings and contradictory of all common sense—will die as hard as any other religion ... but die it must, once the VLBI test of SA is made and reveals SRT for the beautiful half-truth (hence the half-betrayal of both beauty and truth) it is. Others have faith in the theory ... I have faith in the observational test: covariance and spacetime symmetry will be refuted, invariance will be confirmed.

Why am I so serenely confident of the outcome of any honest observational test of the SA issue? Let me confess: Einstein and the string theorists have brain-washed me; they have made of me a lobotomized believer in beauty, damn the torpedoes. But beauty, as remarked in my Preface, is in the eye of the beholder. It just happens that I see invariance outshining covariance as the Sun outshines the Moon. In the beauty sweepstakes, covariance is to me practically a non-starter. She is the girl bespectacled, who by me will never get her neck tackled. She is exactly right for the learned myopes who run physics today—in the grand scheme of things they were meant for each other. For my money, *invariance* is the very name and spelling of beauty.

So, here is my challenge to the relativists: You have made your SA prediction ... dare to test it. If the SA test by the VLBI system supports SRT, then everything I have said here rebounds against me with double force ... and I shall have earned a full measure of retribution through having begged hard for it. As for the relativists, it is not clear what they think they could have to lose by such a test. If it succeeds in showing SRT's predicted departure from classical Bradley SA, they add to their laurels and reconfirm their truths. If not, they can do as they have done in the past in such situations—question the competence of the experimentalists, muddy the waters, even "modify" the theory to give it another crutch. Needless to say, the fact that a charge of experimental incompetence will be their first resort acts as one of the many unspoken deterrents to experiment. The dominance of such social inhibitions renders the "scientific method" in modern practice laughable. For that reason there could arise no occasion to refer to it in this book except for comic relief. The Soviets pioneered politically correct science, but the Americans have honed it to a cutting edge.

References for Chapter 4

[4.1]　J. Bradley, *Proc. Roy. Soc. London* **35**, 308-321 (1728).

[4.2]　P.G. Bergmann, *Introduction to the Theory of Relativity* (Prentice-Hall, New York, 1946).

[4.3]　G. B. Airy, *Proc. Roy. Soc. London* **20**, 35-39 (1871).

[4.4]　J. Aharoni, *The Special Theory of Relativity* (Oxford, London, 1965), 2nd ed.

[4.5]　C. Møller, *The Theory of Relativity* (Oxford, London, 1972), 2nd ed.

[4.6]　A. Einstein, *Ann. Phys.* **17**, 891-921 (1905); variously reprinted.

[4.7]　B. O. Peirce, *A Short Table of Integrals* (Ginn, Boston, 1929), #598.

[4.8]　T. E. Phipps, Jr., *Phys. Essays* **4**, 368-372 (1991).

[4.9]　J. L. Synge, *Relativity: the Special Theory* (North-Holland, Amsterdam, 1965), 2nd ed.

[4.10]　P. Naur, *Phys. Essays* **12**, 358-367 (1999).

[4.11]　G. Herglotz, *Ann. Phys.* (Leipzig) **31**, 393 (1910); F. Noether, *Ann. Phys.* (Leipzig) **31**, 919 (1910).

[4.12]　P. W. Bridgman, *The Logic of Modern Physics* (MacMillan, New York, 1927).

No man hath certainly known, nor shall certainly know, that which he saith about the gods and about all things; for, be that which he saith ever so perfect, yet does he not know it; all things are matters of opinion.

—Xenophanes of Kolophon, *Fragments*

"We are not sure of many things and those are not so."

—Oliver Wendell Holmes (in correspondence)

Chapter 5

Electrodynamic Force Laws

5.1 Electromagnetic force in SRT

In traditional physics we find at work two separate, parallel, and to some extent complementary themes drawn from mechanics: Force and energy. In treating electromagnetism, the force approach fits with field theory—in the sense that the $\bar{E}$-field, for instance, is defined as electric *force* on unit charge. The alternative energy approach is much used in quantum mechanics, since that discipline is normally based on a Hamiltonian (energy) method. This produces a tension, in that effects such as that of Aharonov-Bohm[5.1] seem to favor physicality of the (energy) potentials, whereas field-theoretical doctrine insists that only the fields (forces) have physical meaning. Maxwell himself never entered this debate, since his formulation[5.2] employed potentials and did not represent the modern exclusively *field* form of "field theory" (which was due to Heaviside and others). In "conservative" situations the two can generally be reconciled by expressing forces (fields) as derivatives of potentials. (A physical situation is conservative if idealized in such a way as to banish externals as irrelevant—in which case energy has no place to come from or go to, so it must be "conserved.") Thus in terms of an electric scalar potential ϕ and a magnetic vector potential $\vec{A}$ the electric and magnetic fields in Maxwellian field theory are represented[1.1] as

$$\vec{E}_{Max} = -\vec{\nabla}\phi - \frac{1}{c}\frac{\partial \vec{A}}{\partial t}, \qquad \vec{B}_{Max} = \vec{\nabla}\times\vec{A}. \qquad (5.1a,b)$$

In view of the Aharonov-Bohm evidence[5.1], the question remains open, whether the fields or the potentials are more fundamental in physics. The issue is often settled by doctrinaire pronouncements in favor of fields, usually bolstered by unanswerable *dicta* about "gauge invariance." Like spacetime symmetry, gauge invariance is either trivially obvious or somewhat dubious. We shall not go into that here, but advise keeping an open mind.

The Maxwell field equations, (1.1), define (with the help of boundary conditions) an electric-field solution $\vec{E}_{Max}$ that seemingly ought—since "charge" is electric and since electric field is force on unit charge—to tell us the force on unit charge. But no ... it turns out that, although motion is "relative" (so that "moving" and "static" are not invariant concepts, hence not "meaningful"), moving charge is different from static charge, so our definition of "force on charge" is in need of redefinition. Sure enough then, in conventional electromagnetic theory we solemnly march up this same hill once more and concoct a brand-new *postulate*, a Lorentz force law, to treat the case of force on moving charge,

$$\vec{F}_{Lor} = q\left(\vec{E}_{Max} + \frac{\vec{v}}{c}\times\vec{B}_{Max} \right), \qquad (5.2)$$

which expresses the electromagnetic force actually exerted on electric charge q. Here $\vec{v}$ is the velocity of charge q in the observer's laboratory or inertial system, and the field quantities are as measured at the observer's (stationary) field point—where also the force is exerted and the charge is instantaneously located. Since the moving test charge q here acts as a sensor of the field, it is apparent that the "$\vec{v}$" in (5.2) is the same parameter as our previous detector velocity $\vec{v}_d$.

Notice, then, that the charge moves *through* the field point with velocity $\vec{v} = \vec{v}_d$. Thus, finally, *finally*, in (5.2) field theorists force themselves to recognize that charges *can* move with respect to the field point. Electric charge (acting as sensor or "field detector") is not, after all, absolutely fixed there for all time, as Maxwell's equations plainly have it. (It takes a postulate to break a postulate—the irresistible force that moves the immovable object.

During 150 years, no mathematician or logician has objected to postulates contradicting postulates in electromagnetic field theory or elsewhere in physics, as far as I can discern.) And when that relative motion of charge occurs—what do you know?—something physically new happens: a "magnetic" action on "electric" charge … resulting just from relative motion! The wonders of the axiomatic method will never cease.

To reiterate, the Maxwell field equations do *not* allow sinks to move with respect to field points—those equations not being parameterized to allow such motions. (That non-parameterization is precisely and solely what gives rise to spacetime symmetry.) In previous chapters we have condemned this as a manifest shortcoming of established field theory. Suppose, as advocated there, we had chosen from the start an invariant (Hertzian or neo-Hertzian) formulation of the field equations themselves, such that sink (or "test charge") motions with respect to field points are allowed *ab initio* through being described by the "detector velocity" parameter $\vec{v}_d$. Then might we not anticipate that a law of force on electric charge analogous to (5.2) would "fall out" automatically from such adequately parameterized field equations, without need for extra postulation? Would that be too much to hope for? Let's look into this.

5.2 Neo-Hertzian force law

In neo-Hertzian electromagnetism our standard procedure is to replace Galilean non-invariant expressions such as (5.1) formally with their higher-order invariant counterparts. Thus, invariant forms of the field quantities, which we have termed neo-Hertzian, can be expressed in terms of potentials as

$$\vec{E}_{inv} = -\vec{\nabla}\phi - \frac{1}{c}\frac{d\vec{A}}{d\tau_d} \; , \qquad \vec{B}_{inv} = \vec{\nabla} \times \vec{A}. \qquad (5.3a,b)$$

These are consistent with and derivable from the neo-Hertzian field equations (3.9). Thus the second follows from (3.9c), since the vanishing divergence of $\vec{B} = \vec{B}_{Hz} = \vec{B}_{inv}$ implies that this quantity can be expressed as the curl of some vector field, and the first follows from putting (5.3b) into (3.9b), with $\vec{E} = \vec{E}_{Hz} = \vec{E}_{inv}$ and $\tau = \tau_d$, to yield

$$\vec{\nabla} \times \vec{E}_{inv} + \frac{1}{c}\frac{d}{d\tau_d}\vec{\nabla} \times \vec{A} = \vec{\nabla} \times \left(\vec{E}_{inv} + \frac{1}{c}\frac{d\vec{A}}{d\tau_d} \right) = 0 \,.$$

This vanishing of a curl indicates that the bracketed expression can be equated to the gradient of some scalar function, which implies (5.3a). The fields will then transform invariantly,

$$\vec{E}'_{inv} = \vec{E}_{inv}, \qquad \vec{B}'_{inv} = \vec{B}_{inv}, \qquad (5.4)$$

under the neo-Galilean inertial transformations [Eqs. (3.20), (3.22), (3.23)] provided the potentials also transform invariantly,

$$\phi' = \phi, \qquad \vec{A}' = \vec{A}. \qquad (5.5)$$

Unfortunately, an invariant form such as (5.3) is not directly useful to the laboratory worker, since it quantifies time by readings of a clock co-moving with the test particle ("detector")—something not readily verified in practice. The situation is precisely the same here as it was for the mechanics of particle motions, Chapter 4, Section 8. We must recognize, in order to translate from the invariant-force form to the more useful laboratory-force form, that the $\vec{E}, \vec{B}$ fields are *forces* (per unit charge) and that they transform according to the rules laid down in Eqs. (4.27) and (4.29) for mechanical forces—by extrapolation construed as *rules for all forces*.

The first step in this translation is to separate the forces into spacelike and timelike parts. The $\vec{B}$-field is easy, since it is purely spacelike [as (5.3b) shows], and so is directly invariant according to the rule (4.29). That is,

$$\vec{B}_{Hz} = \vec{B}_{inv} = \vec{B}^{(spacelike)}_{inv} = \vec{B}^{(spacelike)}_{lab} = \vec{B}_{Max} = \vec{\nabla} \times \vec{A}. \qquad (5.6)$$

There is nothing new here. But (5.3a) requires us to recognize that we are attributing to $\vec{E}_{inv}$ a mixed character, as the sum of an invariant spacelike part obeying (4.29),

$$\vec{E}^{(spacelike)}_{inv} = \vec{E}^{(spacelike)}_{lab} = -\vec{\nabla}\phi, \qquad (5.7)$$

which resembles a static gravitational potential (being without time dependence), and a timelike or "magnetic" part, dependent on motion of the test charge, *viz.*, from (5.3a),

$$\vec{E}^{(timelike)}_{inv} = -\frac{1}{c}\frac{d\vec{A}}{d\tau_d} = -\frac{1}{c}\gamma_d\frac{d\vec{A}}{dt}, \qquad (5.8)$$

where t is laboratory frame time and use has been made of Eq. (3.5). This is a force on unit charge obedient to the rule (4.27), which reflects the different clock rates linked by γ_d; namely,

$$\vec{E}^{(timelike)}_{inv} = \gamma_d \vec{E}^{(timelike)}_{lab}, \qquad (5.9)$$

whence, with the aid of (1.8), the laboratory-observable magnetic part of the $\vec{E}$-field is seen (by cancellation of γ_d factors) to be

$$\vec{E}_{lab}^{(timelike)} = -\frac{1}{c}\frac{d\vec{A}}{dt} = -\frac{1}{c}\left(\frac{\partial\vec{A}}{\partial t} + \left(\vec{v}_d \cdot \vec{\nabla}\right)\vec{A}\right). \tag{5.10}$$

The total force on unit charge, or *electric field*, observable in the laboratory, is the sum of timelike and spacelike contributions, (5.7) and (5.10),

$$\vec{E}_{lab} = \vec{E}_{lab}^{(spacelike)} + \vec{E}_{lab}^{(timelike)} = -\vec{\nabla}\phi - \frac{1}{c}\frac{d\vec{A}}{dt}$$

$$= -\vec{\nabla}\phi - \frac{1}{c}\left(\frac{\partial\vec{A}}{\partial t} + \frac{1}{c}\left(\vec{v}_d \cdot \vec{\nabla}\right)\vec{A}\right). \tag{5.11}$$

The fact that "electric field" is thus represented as a sum of completely disparate contributions, timelike and spacelike, suggests that field theory may be concealing more than it reveals. That is, the physical mechanisms behind these two parts could be completely different (*e.g.*, spacelike possibly *via* acausal action of virtual particles, timelike *via* causal real particles—uncompleted *vs.* completed quantum processes). But it fits the spirit of our times to view any successful glossing-over of physical aspects as proof of the power and insight afforded by mathematics. Thus it is mistakenly represented that mathematical understanding can substitute for physical understanding. Those who think in this way will still be propagating the mistakes of their ancestors a thousand years hence, since their mode of appreciating nature lacks external corrective mechanisms and admits of no openings for objective progress.

Some massaging can put (5.11) into a more recognizable form, as follows: A standard identity[2.3] for vector fields [*i.e.*, vector functions of (x,y,z,t)],

$$\vec{\nabla}\left(\vec{a}\cdot\vec{b}\right) = \left(\vec{a}\cdot\vec{\nabla}\right)\vec{b} + \left(\vec{b}\cdot\vec{\nabla}\right)\vec{a} + \vec{a}\times\left(\vec{\nabla}\times\vec{b}\right) + \vec{b}\times\left(\vec{\nabla}\times\vec{a}\right), \tag{5.12}$$

when specialized to $\vec{a} = \vec{v}(t)$ and $\vec{b} = \vec{A}$, yields the simpler identity

$$\left(\vec{v}\cdot\vec{\nabla}\right)\vec{A} = \vec{\nabla}\left(\vec{v}\cdot\vec{A}\right) - \vec{v}\times\left(\vec{\nabla}\times\vec{A}\right). \tag{5.13}$$

[Note that the Lagrangian form $\vec{v} = \vec{v}(t)$ here is not a vector field, hence not subject to spacetime symmetrical analysis.] With the help of (5.13), (5.11) becomes

$$\vec{E}_{lab} = -\vec{\nabla}\phi - \frac{1}{c}\left(\frac{\partial\vec{A}}{\partial t} + \vec{\nabla}\left(\vec{v}_d \cdot \vec{A}\right) - \vec{v}_d \times \left(\vec{\nabla} \times \vec{A}\right)\right). \qquad (5.14)$$

With (5.6) this yields

$$\vec{E}_{lab} = -\vec{\nabla}\phi - \frac{1}{c}\frac{\partial\vec{A}}{\partial t} + \frac{\vec{v}_d}{c} \times \vec{B}_{Max} - \frac{1}{c}\vec{\nabla}\left(\vec{v}_d \cdot \vec{A}\right). \qquad (5.15)$$

The last term here, being the gradient of a scalar quantity, integrates to zero around any closed circuit. Since, according to Maxwell, current flows only in closed circuits, it is apparent that this term would be difficult to observe under normal laboratory conditions. Nothing of the sort has ever been reported as observed. Since current actually does flow (at least sporadically) in open circuits—antennas, plasmas, etc.—it is not impossible that experiments might be devised to test this last term. However, in lack of positive evidence, it might be considered justifiable to drop the last term in (5.15) as (tentatively) unobservable. We are then left with

$$\vec{E}_{lab} = -\vec{\nabla}\phi - \frac{1}{c}\frac{\partial\vec{A}}{\partial t} + \frac{\vec{v}_d}{c} \times \vec{B}_{Max}. \qquad (5.16)$$

An advantage of dropping the extra term is that it permits a Lagrangian formulation, $L = T - U$, where $U = q\phi - (q/c)\vec{A}\cdot\vec{v}$ is a Lagrangian "generalized potential" without physical significance but having the formal virtue of allowing laboratory-observable force to be represented as "generalized force" in the Lagrangian format; e.g., for the x-component,

$$F_x = -\frac{\partial U}{\partial x} + \frac{d}{dt}\frac{\partial U}{\partial v_x}.$$

This possibility is lost if the extra (last) term in (5.15) is retained. Like other formal amenities, however, such as covariance, this one does not speak directly to the physics ... as (I insist) invariance does. (For an excellent summary on use of the Lagrangian method in electrodynamics, see Goldstein.[5.3] My personal bias is against Lagrangian methods, except where they happen to work. This is because they stand or fall *in toto* upon our ability or lack of it to *contrive* a "generalized potential" U having no recognized connection with measurable aspects of physics, but simply conforming to a formal rule. See, however, the discussion of a relationship of the Lagrangian to "action" in Chapter 8.)

Now suppose that instead of postulating a separate force law for force acting on charge q, as is done in Maxwellian electrodynamics, we recognize that the neo-Hertzian electric field in and of itself suffices to describe the entire electromagnetic force on arbitrary electric charge—moving or not. That is, $\vec{E}_{lab}$ in either (5.15) or (5.16) represents *by definition* the laboratory-observable force on unit test charge. Considering (5.16) to describe what is observable in the laboratory as force on single unit charge, the force on a charge of q units is naturally understood to be q times greater; *i.e.*,

$$\vec{F}_{lab} = q\vec{E}_{lab} = q\left(-\vec{\nabla}\phi - \frac{1}{c}\frac{\partial\vec{A}}{\partial t} + \frac{\vec{v}_d}{c}\times\vec{B}_{Max} \right)$$

$$= q\left(\vec{E}_{Max} + \frac{\vec{v}_d}{c}\times\vec{B}_{Max} \right) = \vec{F}_{Lor},$$

(5.17)

where use has been made of the Maxwell $\vec{E}$-field definition, Eq. (5.1a). In this case we see that our neo-Hertzian force law, (5.17), and the Lorentz force law, (5.2), are identical. So, our present theorizing has nothing new to offer experimentalists, apart from whatever can be made of the extra gradient term in (5.15), which generalizes (5.17) to

$$\vec{F}_{lab} = q\vec{E}_{lab} = q\left(\vec{E}_{Max} + \frac{\vec{v}_d}{c}\times\vec{B}_{Max} - \frac{1}{c}\vec{\nabla}\left(\vec{v}_d\cdot\vec{A}\right) \right)$$

$$= \vec{F}_{Lor} - \frac{q}{c}\vec{\nabla}\left(\vec{v}_d\cdot\vec{A}\right).$$

(5.18)

It is to be emphasized that this lab-observable total force on electric charge q is strictly electric—yet it includes the "magnetic" effects associated in traditional theory with charge motion. The physics is in the motion of charge ... forget a separate concept of *magnetism*. Ampère knew this. Heaviside took a step back from it by insisting on distinct electric and magnetic fields.

Since the extra (non-Lorentz) force term in (5.18) exerts no observable action on closed-circuit currents, testing this extra term requires experimental conditions in which current does not flow in closed loops. In this connection it is worth mentioning that anomalous diffusion rates in plasmas, which were at one time in the past reported, might conceivably be accounted for by means of the non-Lorentz term in (5.18), applied to the description of evanescent currents due to fluctuating charge distributions. Random currents of this kind do not in general flow in closed loops. Such observational confirmation would be most gratifying from the standpoint of validating a preference for invariance over covariance. Since I have no access to empirical data on plasma diffusion characteristics (and lack the needed computational resources), such a possible validation lies beyond my personal capabilities. I must therefore leave open the question of observational support for a modification of the Lorentz force law based on Hertzian invariant field theory.

It may be of some interest in this connection to note from (5.18) that the force observable in the laboratory depends on the sum of gradients of two scalar quantities. Thus a sort of superpotential $\psi \equiv \phi + \vec{v}_d \cdot \vec{A}$ can be defined, such that the associated force term takes the form $-q\vec{\nabla}\psi = -\vec{\nabla}\left(q\phi + \vec{I}_d \cdot \vec{A}\right)$, where $\vec{I}_d \equiv q\vec{v}_d$ is the current represented by motion of the detector or test charge q in the observer's inertial frame. The presence of the extra $\vec{A}$-term here might be revealed by experiments conducted in the vicinity of long or toroidal solenoids. Conceivably, the Aharonov-Bohm experiment may qualify as an example, despite the severe muddying of the interpretational waters that has taken place.

At the moment of writing I have no reason to doubt that a full and correct field-theoretical description of physical electrodynamic force, if such exists, would include all terms indicated in (5.18). [Recall that the physics is not constrained by any *a priori* demands of symmetry or covariance.] If so, then (5.18) is a more satisfactory expression for electrodynamic force than (5.17). Eq. (5.18) may be compactly written [in accordance with (5.11)] as

$$\vec{F}_{lab} = q\vec{E}_{lab} = q\left(-\vec{\nabla}\phi - \frac{1}{c}\frac{d\vec{A}}{dt}\right)$$

$$= q\left(-\vec{\nabla}\phi - \frac{1}{c}\left[\frac{\partial\vec{A}}{\partial t} + \left(\vec{v}_d \cdot \vec{\nabla}\right)\vec{A}\right]\right).$$

$$(5.19a)$$

To be sure, this explicitly destroys Lorentz covariance, but here we choose to view that as no loss. Physicists have come to welcome covariance as an invaluable *dictat* that enables them to limit the circle of their thoughts. The result is that they have closed their minds to everything outside that circle. The question—let it trouble such people or not—is whether Mother Nature's mind is equally closed.

In connection with (5.19a) it is of interest to note that the "magnetic" part of the force observable in the lab, namely, $-(q/c)d\vec{A}/dt$, is formally analogous to the (total) time rate of change of particle momentum in mechanics. In fact, Dirac based his quantum theory of the electron on the generalization of mechanical momentum $\vec{p}$ to

$$\vec{p} \to \vec{p} + \frac{e}{c}\vec{A}, \tag{5.19b}$$

so that $(e/c)\vec{A}$, or in our notation $-(q/c)\vec{A}$, is treated as an "electromagnetic momentum," additive to the mechanical $\vec{p}$. This makes sense, of course, only if one is taking total time derivatives, $\vec{F}_{lab} = d\vec{p}/dt$, not partial ones.

To wind up matters of formalism, we may mention that a manifestly "invariant" form

$$\frac{d}{d\tau} = \frac{\partial}{\partial\tau} + \left(\vec{V} \cdot \vec{\nabla}\right) \tag{5.20}$$

is valid, analogous to Eq. (1.8). (Here $\vec{V} = d\vec{r}/d\tau = \gamma\vec{v}$.) The proof uses (3.4), (3.5), and (1.8),

$$\frac{d}{d\tau} = \gamma\frac{d}{dt} = \gamma\left(\frac{\partial}{\partial t} + \left(\vec{v}\cdot\vec{\nabla}\right)\right) = \gamma\frac{\partial}{\partial t} + \left(\gamma\vec{v}\cdot\vec{\nabla}\right) = \gamma\frac{\partial}{\partial t} + \left(\vec{V}\cdot\vec{\nabla}\right).$$

On comparing this with (5.20) we observe that

$$\frac{\partial}{\partial\tau} = \gamma\frac{\partial}{\partial t}, \tag{5.21}$$

a formal analog of (3.5). Eq. (5.20) allows (5.3a) to be written optionally as

$$\vec{E}_{inv} = -\vec{\nabla}\phi - \frac{1}{c}\left(\frac{\partial\vec{A}}{\partial\tau_d} + \left(\vec{V}_d \cdot \vec{\nabla}\right)\right), \tag{5.22}$$

which could be further re-expressed by applying the identity (5.13). Similarly, not only do we have $\vec{F}_{lab} = q\vec{E}_{lab}$ from (5.17) or (5.18), but also

$$\vec{F}_{inv} = q\vec{E}_{inv} = q\left(-\vec{\nabla}\phi - \frac{1}{c}\frac{d\vec{A}}{d\tau_d}\right) = \vec{F}_{inv}^{(spacelike)} + \vec{F}_{inv}^{(timelike)} \tag{5.23}$$

$$= q\vec{E}_{inv}^{(spacelike)} + q\vec{E}_{inv}^{(timelike)},$$

and so on … such manipulations being of primarily formal interest.

Apart from speculation about a possible extra force term, then, the upshot is that we have come back full circle to what was known, the Lorentz force law, (5.17). Does this mean we have gained nothing? No, I think we have gained something of beauty; namely, a set of field equations that not only provide an invariant covering theory of Maxwell's equations, but that describe an *electric* field vector comprehending the full force $\vec{F}_{inv} = q\vec{E}_{inv}$, or $\vec{F}_{lab} = q\vec{E}_{lab}$, acting on *electric* charge q (whether moving or not moving with respect to the field point). This direct linking of electric force action to electric charge, with elimination of "magnetic force" as such, improves the coherence of the physical concept of "electricity." It also improves the *logical economy* of electrodynamics by reducing the postulate count—inasmuch as Maxwell-Einstein theory necessarily treats the Lorentz force law (which allows the test particle to move with respect to the field point) not as a deduction from the Maxwell field equations (which deny such motion) but as an additional postulate. Since the additional postulate permits exactly those test charge motions that were forbidden by the field equations, this new member of the postulate set contradicts the old ones. I find this an example of ugliness within established physics; but have already remarked that beauty is in the eye of the beholder. In a long lifetime I have found few beholders—none in the profession of physics—who share my aesthetics in this matter. The firmly established norm seems to be to make postulates and then to make more postulates that contradict those already made.

5.3 Evidence of the Marinov motor

Concerning the extra non-Lorentz force term in (5.18), it should be mentioned that its confirmation would drive the last nail in the coffin of conventional field theory, as it asserts that vector potential, independently of "field," can exert force. (This was also the Aharonov-Bohm claim—unfortunately thoroughly obscured by the mixing-in of quantum mechanics.) Regarding empirical support, various papers published by myself and others during 1997-8 in *Apeiron, Infinite Energy* magazine, and elsewhere claimed observational confirmation and semi-quantitative agreement with a theory published by J. P. Wesley showing operability [on the basis of the force $-(q/c)d\vec{A}/dt$ of Eq. (5.19a)] of the so-called "Marinov motor." This used a toroidal magnet enclosing all its flux, which acted upon current flowing in a circuit located in the external region where $\vec{B}$ vanished but $\vec{A}$ did not. Reactive torques upon the magnet were measured that should not have existed according to the Lorentz force law, but were non-vanishing according to (5.18). These torques seemed too large to be attributed to leakage flux.

They showed rough quantitative agreement with the extra force term in (5.18) ... but not so unambiguously as to be decisive. It cannot be said at this time that empirical evidence either for or against the extra force term in (5.18) is conclusive.

The matter deserves to be settled. It is far from academic. In *Apeiron* **5**, No. 3-4, 193-208 (1998) I described my observational results, and in *Infinite Energy*, Issue 19, 62-85 (1998) I made engineering-type calculations showing that a scaled-up Marinov motor promised sufficient torque to allow its application to an electrically-driven automotive "motor in a wheel." I recall building a small, crude demo motor (never reported), a multi-turn brushless design requiring switching (current reversal) each half-turn, which spun very nicely but was not self-starting in all orientations. Several independent non-academic investigators—notably Tom Ligon and J. D. Kooistra—got positive Marinov motor results, but none was permitted to publish in a first-line journal, so the claims, obscurely reported, were ignored. In a more healthy condition of American science the mere rumor of a new electromagnetic force would stimulate widespread response in a dozen laboratories, including the electrical measurements division of NIST, the government agency that has the taxpayer-mandated "mission" to investigate such things.

5.4 Other electrodynamic force laws

Over many years diverse empirical evidence has accumulated that the Lorentz force law may not altogether cut the mustard. (If this is true, the same may well have to be said eventually of field theory itself.) Such evidence skulks in the shadows, owing to a steadfast determination of the physics establishment that none of it shall see daylight. (Comparable zeal directed instead to the satisfying of curiosity might prove gratifyingly effective in producing tangible progress.) To document this impression thoroughly would require a lifetime of scholarly effort, and would doubtless encounter barriers to publication from the usual suspects. Plenty of obscurely-published evidence can be found that will be dismissed as anecdotal. As originally intended, this work would have sampled it; but this has had to be curtailed owing to the author's health problems. A little digging in the voluminous shadow literature for "Ampère longitudinal forces" will turn up much of it.

Let me preface what follows by interpolating a few more of my general remarks about science. In modeling *science as process* two antithetical extremes suggest themselves: that of an inexorable tide of unidirectional progress, irreversible and ever-blessed by a consensus of the righteous; and that of a blind bat in a bottle. The factual and historical situation, it seems to me, partakes of both—perhaps about equally. Marchers in the army of science itself acknowledge only the first of these models, but historians of science must surely awake on bad nights in a sweaty suspicion of the second.

As a prime application of the blind bat model, I may point to the handling of Newton's third law (equality of magnitude, oppositeness of direction, and collinearity of all forces of action-reaction of any kind between any two force centers) in modern physics. SRT's imperious demand for "universal covariance" excludes consideration of any force law but that of Lorentz, which violates Newton's third law. Such violation (implying "bootstrap-lifting," the spontaneous abrogation of momentum conservation) has never been *observed* in the real world ... but that world has become so totally subordinate in the minds of theorists to the "world" of SRT that such a lack of empirical confirmation merits little attention in today's textbooks ... which prefer to dazzle the reader's eye with the sheer number of SRT's empirical *confirmations*. It should be a recognized principle of jurisprudence that unlimited increases in the quantity of irrelevant evidence cannot lend it weight.

If we go back in history to the 1820's, when electricity was first being explored scientifically, we find Ampère, the venerated father of the subject, doing a series of exemplary null experiments that systematically pinned-down the law of force action between pairs of "current elements." As a demonstration of the power of abstract reasoning, nonlinearly enhanced by cunning observation, and as a paradigm of the fugal harmony attainable by physics in the hands of a master of the contrapuntal use of theory and experiment, Ampère's work, culminating in his "law," has never been excelled. Its pedagogical value to later generations of students would be inestimable. (The interested reader can find out more from the books of Peter and Neal Graneau.[5.4-5.6])

So ... what actually happened? History shows Ampère's original law being squelched and his work forgotten or marginalized. The reason? Again, SRT—the great destroyer not only of

concepts but of history—has been at work. Ampère's law embod-
ies Newton's third law. (Ampère found that the purely empirical
means accessible to him were inadequate to pin down the EM action-reaction law completely; they had to be supplemented by the
assumption of Newton's third law.) Thus Ampère falls afoul of the
new enlightenment ("progress") that rejects Newton's third law.
Therefore, Ampère's law in all modern textbooks has been not
merely soft-pedaled but totally omitted. For today's student—
presto—it does not exist ... down the memory tube. Instead,
something *called* "Ampère's law"—a politically correct *Ersatz* involving a $\vec{B}$-field Ampère never used (not being a field theorist)—appears in the textbooks to honor his name by abusing it.
This, despite the fact that the politically uncorrected Maxwell in
his *Treatise*[5.2] terms Ampère's original, non-*Ersatz* law "the cardinal law of electro-dynamics."

What, then, is this cardinal law? Allow me to set it down
here reverently, in the likely expectation that it has never before
met the reader's eye. It expresses the force action of a vector current element $I_1 d\vec{s}_1$ upon another element $I_2 d\vec{s}_2$ as

$$\vec{F}_{12}^{(Ampère)} = \vec{r}\,\frac{I_1 I_2}{r^3}\left(\frac{3}{r^2}\left(d\vec{s}_1 \cdot \vec{r}\right)\left(d\vec{s}_2 \cdot \vec{r}\right) - 2\left(d\vec{s}_1 \cdot d\vec{s}_2\right)\right), \qquad (5.24)$$

where $\vec{r} = \vec{r}_2 - \vec{r}_1$ is the vector relative position of the centers of the
two elements, $r = |\vec{r}|$ is their separation, and distance is measured
in centimeters, force in dynes, and current in abamps (1
abamp = 10 Ampères). Positive force denotes repulsion, negative
force, attraction. [If, in conformity with the SI system of units, one
multiplies (5.24) by $\mu_0/4\pi$, distance is expressed in meters, force
in Newtons, and current in Ampères. To do this is in my opinion
a trifle silly, since "meters" are not natural distance units to use in
the laboratory—not in my modest one, anyway—and dynes are
close enough to milligrams to be comparatively natural force
units, conformable to intuition. Moreover, with my failing facul-
ties I find it easier to remember the ratio 10 of abamps to amps
than the value of $\mu_0/4\pi$. It's even typographically simpler!]

When the Lorentz law (5.2) is applied to description of the
same "magnetic" situation—the action of one current element on
another—it yields the law

$$\vec{F}_{12}^{(Lorentz)} = \frac{I_1 I_2}{r^3}\left(d\vec{s}_2\left(d\vec{s}_1 \cdot \vec{r}\right) - \vec{r}\left(d\vec{s}_1 \cdot d\vec{s}_2\right)\right) \qquad (5.25)$$

in units similar to those of (5.24). Observe the dramatic difference between these two: Eq. (5.24) shows force proportional to $\vec{r}$, hence aligned with the vector joining the two element positions, and also shows symmetry under interchange of subscripts 1 and 2. Consequently, Newton's third law,

$$\vec{F}_{12}^{(Ampère)} = -\vec{F}_{21}^{(Ampère)}, \tag{5.26}$$

is obeyed by (5.24), since $\vec{r} = \vec{r}_2 - \vec{r}_1 \rightarrow \vec{r}_1 - \vec{r}_2 = -\vec{r}$. In contrast, (5.25) has no such symmetry. The force exerted by element 1 on element 2 bears no relation to that exerted by 2 on 1, as to either magnitude or direction. It is thus left as a discussion topic to explain why no electrodynamic force imbalances are observed in nature. There is a theorem bearing on this that explains why closed current loops exhibit no force imbalances; but in broader physical contexts the question remains open … as does the question of the status of momentum conservation, once its classical mechanical basis in Newton's third law has been discarded. It also remains to be answered why, if there exist forces that violate Newton's third law, there also exist forces that *do* obey it. The *Herrgott* seems lacking in the consistency little minds might desire of Him.

Historically, (5.25) grew out of an objection by Grassmann[5.7] to the Ampèrian idealization of "current elements" as mathematical points. Grassmann viewed it as illegitimate to shrink a directed current element to a point—an opinion that implicitly specifies a permissible order of limiting processes in the application of mathematics to physical approximation. In effect, his contention was that in the limit of "element" definition the full effect of current directedness had to survive; consequently Ampère's deductions were null and void … and Newton's third law inapplicable. Whittaker[5.8] (page 86) seconds Grassmann's judgment in these terms: "The weakness of Ampère's work evidently lies in the assumption that the force is directed along the line joining the two elements; for in the analogous case of the action between two magnetic molecules, we know that the force is *not* directed along the line joining the molecules." Since he gives no reference, Whittaker fails to make it clear how he has become privy to the congress between two magnetic molecules. ("Vas you dere, Chollie?") But I daresay if the evidence is empirical it bears on molecules in fairly close proximity, so that molecular dimensions are not negligible compared to their separations. Thus it can have no bearing on the mathematical question of order of limiting processes (element length *vs.* element separation) crucial to settling the legitimacy of Ampère's idealization of neglecting element orientation in approximating element interaction.

Let us detour a moment to contemplate Grassmann's Eq. (5.25) in abstract terms. I submit that as physics this equation is foolish on the face of it. For it ascribes to current elements an asymmetry for which no conceivable inherent physical basis can exist. In nature element 1 possesses no special "oneness" and element 2 no special "twoness." Each partakes equally of oneness and twoness. It can make no physical difference which number is assigned to which. Hence it is desirable that an adequate mathematical representation of their interaction exhibit a corresponding interchangeability, *i.e.*, explicit symmetry between 1 and 2. Theorists who are daft for spacetime symmetry, a highly non-obvious symmetry, glibly sacrifice to it action-reaction (1-2) symmetry, a comparatively obvious one. It is like human sacrifice on the altar of a holy goat. Modern theoretical physics in its ability to lord the subtle over the sensible has evolved from the rabbinical to the terminally talmudic. I set out in this section to illustrate the usefulness of the blind bat model of scientific process. Is this picture beginning to gain some color of justification?

Why was (5.24) once venerated as the "cardinal law"? Because to this day, as in Maxwell's time, it has never been contradicted by any observation. If you truly desire to know what you will observe in your own laboratory, and are curious enough about the subject to abandon ideology, use that law. On it you can rely. Is it not strange, then, that today's physics textbooks make no mention of it? Perhaps not so strange, if you reflect upon the authors' state of mind: they write not about physics but about truth as they know it and desire it to be known to posterity. They do not wish to pollute the Pierian mind of the student with false themes from a bygone era, darkened by the dismal miasma of misunderstanding that preceded Einstein's unique sunburst of enlightenment. He has sounded forth the trumpet that shall never call retreat. Progress has progressed … henceforth, let there be no regress … and no moaning of the bar. This ratcheting process reflects the central principle of pedagogy responsible for such giant steps for mankind as the trumpeted advent of the New Math, to assure that no child be left behind, and the soft-pedaled abandonment of English grammar, to assure that no teacher be left behind.

Numerous "electricians," as Whittaker[5.8] called them, of the nineteenth century concocted their own "improvements" on Ampère's formula (5.24). Grassmann[5.7] led the pack by proposing (5.25), later incorporated into the Lorentz force law as its magnetic part. I am not a sufficient scholar to contribute to this field, so will make no attempt even to tabulate these many mutually-contradictory improvements. Let one suffice as a sample. I once, perhaps mistakenly, attributed[5.9] to Riemann the proposed law

$$\vec{F}_{12}^{(Whittaker)} = \frac{I_1 I_2}{r^3} \left(d\vec{s}_1 \left(d\vec{s}_2 \cdot \vec{r} \right) + d\vec{s}_2 \left(d\vec{s}_1 \cdot \vec{r} \right) - \vec{r} \left(d\vec{s}_1 \cdot d\vec{s}_2 \right) \right), \quad (5.27)$$

and ascribed the attribution to Whittaker. But I find on re-consulting Whittaker[5.8] (page 87) that there is no basis for such an attribution and that (5.27) appears to be Whittaker's own interpolation into or improvement on history. He claims for it the evident merit over the Lorentz (Grassmann) law of expressing explicit 1-2 symmetry, justifying this on the basis[5.7] that "the law of Action and Reaction may not be violated." [Eq. (5.27) is in fact a symmetrization of (5.25).] But he ignores the fact that *Newton's* law of Action and Reaction *is* violated by any force, such as that of (5.27), that is not aligned with the separation vector $\vec{r}$. Ampère knew this and acted on the knowledge. His improvers all have chosen to forget it. Whittaker further overlooks the non-covariance of his (5.27), which puts him as firmly in opposition to the New Wave of relativistic truth as any of the other seekers for alternatives to the Lorentz law. In this, of course, I heartily second him on the side of theory. However, there is empirical evidence[5.9;5.10] against (5.27) and for Ampère's (5.24).

5.5 Sick of field theory? ... (the Weber alternative)

And now for something completely different. In 1846 Wilhelm Weber[5.11] devised an instant-action electrodynamics, employing the (relative) separation coordinate $r = |\vec{r}_2 - \vec{r}_1|$ of two charged point particles and its first two time derivatives, that obeyed Newton's third law, reproduced the original Ampère law (5.24) of action-reaction between current elements, and was derivable from a potential. It would be useful for any physicist to know about this, though few do. The inter-charge force law Weber proposed was

$$\vec{F}_{12}^{(Weber)} = -\vec{F}_{21}^{(Weber)} = \frac{q_1 q_2}{r^3}\vec{r}\left(1 - \frac{\dot{r}^2}{2c^2} + \frac{r\ddot{r}}{c^2}\right), \tag{5.28}$$

where, as before, $\vec{r} = \vec{r}_2 - \vec{r}_1$, and the force is that exerted by charge 1 on charge 2. (This is in electrostatic units. To express it in SI [*Stupidité Incroyable*], multiply by $1/4\pi\varepsilon_0$.) The corresponding potential from which this force can be derived is

$$U = \frac{q_1 q_2}{r}\left(1 - \frac{\dot{r}^2}{2c^2}\right). \tag{5.29}$$

That is,

$$\vec{F}_{12}^{(Weber)} = -\vec{\nabla}_r U^{(Weber)} = -\frac{\vec{r}}{r}\frac{d}{dr}U^{(Weber)},$$

where in performing the total differentiation we are on notice that

$$\frac{d\dot{r}}{dr} = \frac{dt}{dr}\frac{d\dot{r}}{dt} = \frac{\ddot{r}}{\dot{r}}.$$

Observe that (5.28) is a law of force between point charges, not between current elements. It is therefore more fundamental than Ampère's law, but is complementary to it in that both treat of instant action at a distance and both honor Newton's third law unreservedly, without fudgings. (In a way, Weber provides a *post-facto* line of justification for Ampère's assumption of Newton's third law; for even Grassmann would agree that the third law has to be obeyed between point charges—from which "directed" current effects are deduced by Weber, still in obedience to the third law.) In fact, Weber's theory was the first and only form of electrodynamics that was truly "relativistic," in the sense that it introduced no extraneous "frames" or "observers" for describing nature, but employed only relative coordinates of the active elements—the point charges themselves. Thus relativity existed in its all-time purest form in Europe in the 1850's, before Faraday with his moving lines of force and Maxwell with his fields contaminated physics with their intangible metaphysical entities—an evolution (away from relativity) from which it may never recover ... since it defined thenceforth the forward direction of *progress*.

Wherein lay the superiority of the field mode of description? Simply in its ability to predict the time delays of causal "propagation." Causes at point A produced later effects at point B, the two being linked by an appearance of something moving from A to B at speed c. (This applies to radiation. Many assume that it applies also to electromagnetic forces, but there is no empirical evidence to back this.) On this slender evidence of appearance, history (which is to say the consensus of authority) awarded field theory total, permanent victory, with no *reclama* allowed from the loser, Newtonian electrodynamics.[5.6] The irony is that even while losing the war, the latter was winning the battle. That is, the European electrodynamicists, adhering to the strictly Newtonian instant-action tradition, were beginning to discover speed-c time delays emerging from their formalism. (This should be no stunning surprise, since c, or the equivalent, made its first appearance in Weber's formulas.)

Thus Kirchhoff in 1857 scored a significant breakthrough.[5.12] The Graneaus[5.6] (on page 41) describe it in these terms:

> Kirchhoff proved with circuit theory that voltage and current waves travel along wires with the velocity of light. This remarkable fact arose from multiple inductive and capacitive far-actions of huge numbers of current and conductor elements. Kirchhoff was in fact the first to derive delays in the transmission of electrical disturbances along conductors with a many-body interaction model. The strange aspect of the complex far-action calculations is that they predict the same time delays as the simple energy transport model of field theory.

For future reference we may note here the central role of the many-body problem, whose far-reaching significance will be dwelt on further in Chapter 8.

Recently a triumphant confirmation of the fruitfulness of the Kirchhoff approach has been achieved by Neal Graneau,[5.13] who made computer calculations, using Ampère's original force law (5.24), to show that instant actions of large numbers of coherent current elements separated by distance D in free space from large numbers of incoherent elements induce a coherent response in the latter that grows in time and that is delayed in onset proportionally to D. (The delay results jointly from inertial sluggishness of the material current elements and from inverse-square weakening of the Ampère force with distance.) These, broadly speaking, are the characteristics of far-zone radiation, as measured, *e.g.*, by antennas. The quantum aspect is missing, but it is missing also from all undoctored forms of field theory. Breathes there a physicist with soul so dead that he feels no stir of excitement at the possibilities thus offered by the prospect of a deeper understanding of the many-body problem?

Thus it is not true that instant-action models are limited to instant-action predictions. It is only politically convenient in the modern era to let it be thought so, in order that field theory may justify its banishing of all rivals. Had a modicum of pluralism been maintained during the twentieth century in the basic conceptions of physics, who knows or can even guess what benefits to mankind might be conveyed by the resulting theoretical options? At the very least there would be less dogmatism in the classroom ... and, given some hints from an instant-action form of electrodynamics, practitioners of physics would be less in denial of the persistent acausality of quantum mechanics than are those currently dominant savants whose sole intellectual resort is to the plodding causality of retarded-action field theory.

There is much more to be said about Weber's electrodynamics. Fortunately, there is an excellent book on the subject by Assis,[5.14] which cannot be recommended too highly. (In describing Weber's coordinates, Assis terms them "relational" rather than "relative," to avoid confusion with Einstein's usage of "relativity.") Weber's relational coordinates are observer independent and frame independent. Thus they fit, to a degree, with some of the conceptions of general relativity, but not of SRT. A problem arises, however, when one seeks to relate them to invariant proper time of the particle, inasmuch as for a pair of particles a pair of proper times would necessarily be involved. We have no clear knowledge about proper time apart from what the Einstein relation (3.1) tells us—and that relation forces introduction of an inertial frame, willy-nilly. Since an inertial frame is also indispensable to Newtonian mechanics, it appears unavoidable to employ the frame formulation of basic physics. If so, Assis shows that Weber's force law (5.28) takes the frame-coordinate form

$$\vec{F}_{12}^{(Weber)} = \vec{r}\,\frac{q_1 q_2}{r^3}\left(1 + \frac{1}{c^2}\left(\vec{v}\cdot\vec{v} - \frac{3}{2}\left(\frac{\vec{r}}{r}\cdot\vec{v}\right)^2 + \vec{r}\cdot\vec{a}\right)\right), \qquad (5.30)$$

where $\vec{r} = \vec{r}_2 - \vec{r}_1$, $\vec{v} = \vec{v}_2 - \vec{v}_1 = d\vec{r}_2/dt - d\vec{r}_1/dt$, $\vec{a} = d\vec{v}_2/dt - d\vec{v}_1/dt$. When we seek an invariant formulation on the time side, difficulty arises through the circumstance that Weber's $\vec{r}$ involves a single time parameter, whereas any two particles in general involve two (proper) time parameters. From this, one cannot avoid the suspicion that Weber's electrodynamics would profit from an overhaul to bring it into the 21st century. In Chapters 6-8 we shall develop a "collective time" mode of description that may prove useful for achieving the necessary higher-order single frame-time parameterization to allow a rebirth of the Weber formulation of electrodynamics.

The subject of an invariant electrodynamics is evidently a work in progress—or would be if any work were being done on it at all. Herein lies a challenge to the coming generations of physical thinkers. The Weber force between a pair of charged particles—as under-developed during the nineteenth century—seemed to exist in a vacuum of concepts, as if remote from a Machian universe of matter. It is only when we try to use this force in an equation of motion that we meet the necessity for a concept of inertia and recognize (a) that mass plays a role, (b) that frames are useful, particularly for many-body description, and (c) that (as Einstein has taught us) time description is not trivial. In this context, Assis's pioneering work[5.14] provides much food for thought.

5.6 Chapter Summary

We have seen that SRT depends on a covariant force formulation requiring the Lorentz force law and allowing no option. Thus believers in SRT not only lack motivation to seek improvement in our basic understanding of electrodynamics, but are committed to resisting any suggestion of the possibility of such improvement. (There is no weapon so effective against the increase of knowledge as an elegant sufficiency of established knowledge.) So deep is their commitment to SRT that they are willing to reject action-reaction symmetry :for the sake of preserving spacetime symmetry, although no direct violation of the former has ever been observed empirically ... and no direct empirical support for the latter has ever been reported. Yet they tell us that SRT is the best-confirmed (empirically!) of all physical theories. With friends like these, O Physics, what need have you of enemies?

The field equations of Maxwell admit of no charge motions with respect to the field point and thus require separate postulation of a "force law" (that of Lorentz) to describe the effect of such charge motions. (Recall that Maxwell's electric field is *defined* as force on a unit charge *at rest* at the field point, never in motion!) This need for a separate force postulate is removed by substituting an invariant formulation of field theory that is adequately parameterized to allow charge ("field detector") motions with respect to the field point. The resulting invariant electrodynamic force law (5.18) is identical to the Lorentz law except for an extra term that is the gradient of a scalar function and thus is not observable by means of closed-loop current interactions. Empirical confirmation of such an extra term would be difficult; it might conceivably be inferred through observation of plasma diffusion-rate anomalies. Claims have been made by unanointed amateurs such as myself, that the "Marinov motor" works—which can only be true if the extra force term $-(q/c)\vec{\nabla}\left(\vec{v}_d \cdot \vec{A}\right)$ has some counterpart in nature. This extra term, needed for invariance, destroys covariance. Its physical validity must be considered empirically undecided at this time ... and for however long Einstein adulation continues to block scientific curiosity.

The ability of the neo-Hertzian field equations to tell us the total force on electric charge q, via $\vec{F} = q\vec{E}_{Hz}$, directly from the $\vec{E}$-field solution of those field equations without extra "force law" postulation, represents a significant advance in logical economy for electromagnetic theory. In fact it amalgamates electrodynamics and electromagnetism, so that the two become the same theory. What does it mean physically that a law of force on moving charge comes automatically out of an invariant formulation of electromagnetic theory? It means that our perception of the "physics" of magnetic force on moving charge was a misapprehension. Such physics was not a product of nature but of our (covariant) way of describing nature. The essentially electric character of all force on electric charge emerges only with an invariant reformulation of field theory.

Ampère, working from premises of Newtonian instant-action, proposed a specific law of force action between current elements that obeys Newton's third law and has never been refuted by observation. No way has been discovered to link this to field theory. Weber, working from the same premises, deduced a more basic law of action between point charges that describes both electric and magnetic effects and agrees with Ampère's law. It failed to describe radiation directly, but achieved (through work of Kirchhoff[5.12]) a capacity to describe time-retarded propagation in conducting wires, and (through work of Graneau[5.13]) the same capacity in respect to free-space propagation of electric action (light). The Weber instant-action descriptive

theme was abandoned too abruptly in favor of field theory to allow full assessment of its potentialities as physics. The subject
may truly be said to be in its infancy. Historically, it was done to
death in a Darwinian struggle. According to the victors in this
struggle it deserved to die. They have written the history, and
only their judgments are known today. It was the same with the
Israelites, who survived to write a Bible, in regard to the Canaan-
ites, who didn't.

In terms of empirical support, Ampère's original law describing the interaction of current elements is the most promising of
numerous nineteenth-century proposals alternative to the Lorentz law. Unfortunately, the position of Ampère's law squarely in
the middle of the Newtonian instant-action tradition makes any
building of a bridge to field theory, whether invariant or covariant, both practically and conceptually very difficult. I know of no
way to get from either unimproved (Maxwellian) field theory or
improved field theory (Hertzian or neo-Hertzian invariant electrodynamics) to any Ampère-type law of force between current
elements or to the Weber law. As far as I can see, all one gets from
invariant field theory is a warming-over of the Lorentz force law,
with possible giblets [Eq. (5.18)] on the side. My personal opinion
is that even the best of field theories—for all their acclaimed
prowess in describing radiation—can never do more than a partial, sketchy job of physical *force* description. Rather than seeking
a linkage whereby the gap between field theory and the empirically supported longitudinal Ampère forces can be bridged, I
should guess it to be more promising to follow the Kirchhoff clue
and look for some altogether new conceptualization of the many-
body problem. The future of fundamental electromagnetic theory
doubtless lies in the direction of effecting an integration with
quantum mechanics. Still, the strongest future physics will require an arsenal of concepts, methods, and presuppositions—a
varied weaponry suitable for attacking problems on multiple lev-
els of sophistication. This is the pluralism I have argued for. I am
obliged to leave the subject as another challenge to later generations of physical thinkers, hypothetically not so impressed by the
truths of earlier generations as to be incapable of unlearning a
generous part of their higher educations.

Finally, I claim to have lent some support to the blind-bat-in-
a-bottle model of theoretical physics as process. The bat keeps
hitting the bottle, an unforeseen and invisible obstacle, and

bouncing back off it, changing direction without ever changing course. In order to stick to its SRT course it abandons (while demonstrating) action-reaction symmetry when it hits its nose against that side of the bottle. Within SRT itself, it develops calluses from multiple impacts, as in the collision with the "rigid body," which necessitates giving up the claim to cover Newtonian rigid-body mechanics, or the collision with group theory, which necessitates redefining "inertial system" to accommodate an *ad hoc* "Thomas precession," made for the occasion ... every check, every turn-about, every nose-bloodying event being hailed and defined as progress. Indeed, on observing the supernatural, the almost divine, persistence of the creature—the monomania with which it sticks to its "self-consistent" course of mad mathematical logic despite all impediments nature can put in its path, the fanaticism with which it crushes all would-be pluralistic attempts to diversify its presuppositions—an objective observer can hardly avoid the inference that the bat is not merely blind but rabid.

References for Chapter 5

[5.1]   Y. Aharonov and D. Bohm, *Phys. Rev.* (Ser. 2) **115**, 485-491, 1959.

[5.2]   J. C. Maxwell, *A Treatise on Electricity and Magnetism* (Dover, New York, 1954), 2 vol., 3rd ed.

[5.3]   H. Goldstein, *Classical Mechanics* (Addison –Wesley, Cambridge, MA, 1950), Section 1-5.

[5.4]   P. Graneau, *Ampere-Neumann Electrodynamics of Metals* (Hadronic Press, Palm Harbor, FL, 1994), 2nd ed.

[5.5]   P. and N. Graneau, *Newton versus Einstein* (Carlton Press, New York, 1993).

[5.6]   P. and N. Graneau, *Newtonian Electrodynamics* (World Scientific, Singapore, 1996).

[5.7]   H. Grassmann, *Ann d. Phys.* lxiv (Vol. **64**), 1-18 (1845).

[5.8]   E. Whittaker, *A History of the Theories of Aether and Electricity* (Harper, New York, 1960), Vol. 1.

[5.9]   T. E. Phipps, Jr., *Hadronic J.* **19**, 273-301 (1996).

[5.10] N. Graneau, T. Phipps, and D. Roscoe, *Eur. Physical J.* D **15**, 87-97 (2001).

[5.11] W. Weber, *Werke* (Springer, Berlin, 1893), Vol. 3, pp. 25-214.

[5.12] G. Kirchhoff, *Poggendorf's Annalen* **100** and **102** (1857).

[5.13] P. and N. Graneau, *In the Grip of the Distant Universe: The Science of Inertia* (World Scientific, Singapore, 2006).

[5.14] A. K. T. Assis, *Weber's Electrodynamics* (Kluwer, Dordrecht, 1994).

> You must accept the truth from whatever source it comes.
>
> —Moses ben Maimon, philosopher (1135-1204)

Chapter 6

Clock Rate Asymmetry

6.1 Distant simultaneity, acausality

In previous chapters I have developed the beginnings of an invariant electrodynamics (wherein length is invariant and the speed of light is not universally constant) as an alternative to Maxwell's electromagnetism. In this chapter I propose to examine an application of such a generalized electrodynamics to revamping some of the basic reasonings underlying Einstein's special relativity theory (SRT) ... in particular, those that close the public, as well as the professional, mind to the possibility of a physically meaningful distant simultaneity.

Standing squarely in the way of the possibility of conceptualizing instant action-at-a-distance, accommodating Newton's third law as physics, accepting quantum non-locality, *etc.*, is Einstein's great insight, the "relativity of simultaneity." With this insight, and its phenomenal, universal sale in the marketplace of ideas, all chance was lost of any subsequent fair hearing for alternatives (of the sort advocated in this book) to the Einstein-Minkowski "world." Even when the facts of quantum experience plainly dismissed that world by exhibiting instant action-at-a-distance as a laboratory phenomenon (experiment of Aspect[6.1], *etc.*), it continued to be an unquestioned major goal of physicists to squeeze quantum mechanics into SRT's Procrustean bed—although the squeezing clearly forced together strange bedfellows. Just how was this unprecedented sale of a counter-factual "fact" accomplished?

This needs to be asked about one of the most successful jobs of radical idea salesmanship ever achieved in the annals of mass mind-closing. So convincing was Einstein's argument as to be truly irresistible, in the sense that nothing known before could withstand it, and everything subsequent has been built upon it. That's quite an intellectual accomplishment, albeit a cruel blighting of preconceptual pluralism, the traditional seedbed of scientific progress. The most deeply ingrained subsequent ills of physics not directly traceable to Maxwell's equations are traceable to the "relativity of simultaneity."

By destroying the conceptual basis for distant simultaneity Einstein got rid of the "now" that each of us *perceives* as dividing past from future. In so doing he discredited perception as a criterion of truth, and removed physics from the realm of personal experience by denying the *description of nature as manifested in experience* as the goal and definition of physics. (The irony is that *agreement with experience* is the hammer invariably used to flatten all opposition to his theories.) Einstein proclaimed that the world as a progression of experiences from past to future was an illusion to be replaced by an invisible web of monolithic worldlines, existing in a Minkowski 4D "world" of spacetime symmetry that was the only reality, wherein *"now"* was physically meaningless. (Subsequently this new reality acquired a curvature that made it really real.) The physical thus became a metaphor for the mathematical, the ultimate repository of truth, and personal experience no longer entered scholarly discussion except to exemplify the snares besetting the unwashed. Philosophers lapped it up. It had just the profundity they had been questing for centuries, and hadn't known since Parmenides (whose doctrine averred that "what is various and mutable, all development, is a delusive phantom"). Physicists lapped it up, not because it tasted good but because all the rest were consuming it in carload lots, and what few competitive ideas there were, such as those of Lorentz, definitely tasted sour. Science popularizers lapped it up because it was grist for their mills. The public lapped it up because it did not do to be left behind by progress, and four-color expositions of it impressed friends when left lying on the coffee table.

So, I ask again, how was this epoch-making sales job accomplished? In a word, through the example of "Einstein's train." That is of course an oversimplification. No single example could achieve such a *bouleversement* of human thought. But it was the clincher. Before it, legitimate doubt could exist—after it doubt became aberration or dementia. Why was the train example so convincing? Simply because its homely materials made it directly accessible to Everyman's thought. Its systematic steps of reasoning so eloquently expressed the triumph of rationality that no sane person could resist. Here was pioneering science brought down to a level any attentive student could follow and appreciate. It was a truly definitive sales approach, the master virtue of which was that it left no reply possible, no objection admissible, no disagreement conceivable. At its conclusion the product had to be bought, regardless of cost to preconceptions.

How, then, comes it that I do not buy it, yet claim to be a rational being? The demented do lay claim to rationality, to be sure, and that is an easy explanation … but, if the reader will do me the favor of sticking with me (which is not unlikely, since he has stuck this far), I shall presently make what I believe to be the beginnings of a *rational* case for the possibility of an alternative self-consistent line of development, featuring the absoluteness of simultaneity, the universality of Newton's third law, and the meaningfulness of *now*. The case for this alternative, like all theorizing, will be far from conclusive, but it should furnish food for thought … and it does submit to crucial testing (*e.g.*, the example of stellar aberration already treated in Chapter 4). If it passes such tests, it will still not be the truth—only a less false mask than the one it replaces. Scientific truth is useful only as a goal … as an attainment it is a plague, a blight, a roadblock to progress. For the scientist it serves the same function as the artificial rabbit in a dog race.

Einstein's conclusion from his train example, reviewed below, is that spatially separated events judged simultaneous by one inertial observer are judged non-simultaneous by another; hence, distant simultaneity is "relative." However, the demonstration of this "fact" depends critically upon a certain model of the light propagation process. That model, due to Maxwell, may or may not be correct physically. We do not know which, because "propagation" by its nature is purely a matter of inference. We cannot know in any ultimate sense of fact what goes on during this non-local process because to find out we would have to terminate (localize) it. This is the old story basic to quantum mechanics: The "propagating" photon is in a quantum pure state, which can be interrogated only by destroying it … and which without interrogation cannot be *known*. The impasse for classical theory induced by this blockage of factual knowledge of "trajectories" is what the wave-particle dualism is about. The propagating mode is the spread-out "wave" one, and the localized-action mode is the "particle" one. The spreading-out is in time as well as space, so an inherent fuzziness mocks any pretend-knowledge (our inferences) about what is where during the main part of the energy-transfer process termed *propagation*. The Maxwell-Einstein inferences differ from other forms of pretend-knowledge about this subject only in sporting by far the most numerous and least imaginative clientele. It is never conservative and seldom safe to treat pretend-knowledge as if it were real.

Permit me, then, to offer below an alternative scheme of inference, based on Hertzian or neo-Hertzian electromagnetism, which will contrast with the Maxwellian scheme but will be in no objectively verifiable sense (logical or phenomenological) either superior or inferior. It will just have a smaller clientele, because it appears to bear the stigma of "acausality." About this, however, we should remind ourselves (a) that photon propagation—the same photon dubbed classically "the field"—far from having anything classical about it, is in physical fact the "most quantum" phenomenon in nature,[1.6] and (b) that quantum mechanics possesses an ineradicably acausal aspect. This latter fact has been labelled "spooky" by Einstein. It is an aspect of the natural world, like some others, in the intelligent design of which neither Maxwell nor Einstein was consulted ... but one to which many of us, less emotionally committed, have latterly managed to accommodate ourselves. The majority of today's physicists have apparently learned to live mentally a double life, in which they accept acausality when wearing their quantum mechanician hats and reject it when wearing their relativist, field theorist, and other more attractively fashionable causal hats.

"Acausality" ought, strictly speaking, to be defined as the occurrence of an effect *before* its cause. It is loose thinking to equate it also to occurrence of an effect at the same time as its cause. If Einstein's view were correct, that the occurrence of two spatially separated events at the "same time" has no meaning, it would be equally *meaningless* to distinguish, or even to attempt to *define*, the concepts "causal" and "acausal"—but today few scientists practice strict thinking on this subject. Bridgman (in *The Logic of Modern Physics*) was among the few outspoken about it:

... I can see no justification whatever for the attitude which refuses on purely *a priori* grounds to accept action at a distance as a possible axiom or ultimate explanation. It is difficult to conceive anything more scientifically bigoted than to postulate that all possible experience conforms to the same type as that with which we are already familiar, and therefore to demand that explanation use only elements familiar in everyday experience. Such an attitude bespeaks an unimaginativeness, a mental obtuseness and obstinacy, which might be expected to have exhausted their pragmatic justification at a lower plane of mental activity.

6.2 Einstein's train on a different track

The considerations that follow will be limited in validity (as far as the Einsteinian analysis goes) to the first order. In Einstein's "train" problem a stationmaster S is at rest at position $x = 0$ at the S-frame ("embankment") time instant $t = 0$ when (a) the midpoint of a train of length $2L$, traveling from left to right at speed v_d, passes him and (b) two simultaneous lightning strikes occur, one at each end of the train. The simultaneity is as judged by S. A train rider R is positioned at the train's midpoint. The lightning flashes are considered to propagate as if in vacuum. We wish to determine how S and R measure or infer the timing of the events of flash arrivals at their eyes (treated as radiation detectors). We shall for simplicity consider everything from the viewpoint of S. Let the propagation speed (phase velocity) of the right-going flash from the *rear* of the train be denoted by v_r, and let the speed of the left-going flash from the *front* be denoted by v_f. According to all theories, given that each flash starts from the same distance L and propagates toward S with the same speed c (*i.e.*, on the assumption that $v_r = v_f = c$), S will see the flashes simultaneously at his (S-frame) time

$$t_1 = \frac{L}{c},\qquad(6.1)$$

since by hypothesis their events of emission were simultaneous in S's view. Such simultaneity in the S-frame is equally valid in Maxwell's and in Hertz's theories, since the light detector (eye of S) is at rest in S, so $v_d = 0$ and the two theories are identical in this special case.

Consider the rear-originating flash, with wave front that moves rightward. It starts at $t = 0$ when R is at $x = 0$. It reaches R and is absorbed there at a later S-frame time t_{rR} and up-track (rightward) position x_{rR}. We have $x_{rR} = v_d t_{rR}$, the distance the train travels at speed v_d in time t_{rR}; and also we have $x_{rR} + L = v_r t_{rR}$, the total distance the right-going flash from the rear of the train travels at light speed v_r in time t_{rR}. (Here we avoid the specialization $v_r = v_f = c$ in order to treat the general case.) Hence, solving these two equations for t_{rR}, x_{rR}, we find

$$t_{rR} = \frac{L}{v_r - v_d} \qquad (6.2a)$$

and

$$x_{rR} = v_d t_{rR} = \frac{v_d L}{v_r - v_d}. \qquad (6.2b)$$

Similarly, let the front-originating (left-going) flash reach R at time t_{fR} and position x_{fR}. We have $x_{fR} = v_d t_{fR}$ for the train travel and $L - x_{fR} = v_l t_{fR}$ for the left-going flash from the front. Hence

$$t_{fR} = \frac{L}{v_f + v_d} \qquad (6.3a)$$

and

$$x_{fR} = v_d t_{fR} = \frac{v_d L}{v_f + v_d}. \qquad (6.3b)$$

These simple and entirely general relations will form the basis for all the discussion that follows.

Case (a). Maxwell-Einstein deductions

Applying his second postulate (of light-speed constancy), based on the Maxwell picture of light propagation, Einstein takes $v_r = v_f = c$, independently of both source motion and detector motion; so Eq. (6.2a) yields, with (6.1),

$$t_{rR} = \frac{L}{c - v_d} = \frac{t_1}{1 - v_d/c} \qquad (6.4a)$$

and

$$x_{rR} = v_d t_{rR} = \frac{v_d L}{c - v_d}, \qquad (6.4b)$$

and Eq. (6.2b) yields

$$t_{fR} = \frac{L}{c + v_d} = \frac{t_1}{1 + v_d/c} \qquad (6.5a)$$

and

$$x_{fR} = \frac{v_d L}{c + v_d}.$$ (6.5b)

By inspection we see that $t_{fR} < t_{rR}$ (actually, $t_{fR} < t_1 < t_{rR}$) and $x_{fR} < x_{rR}$, so the flashes are received non-simultaneously by R, while being received simultaneously by S. This is the basis for deducing the "relativity of simultaneity." (From the non-simultaneity of flash reception events by R, given equality of the propagation distances L in the train system, R infers the non-simultaneity of flash *emission* events—those same events that for S were simultaneous by hypothesis.) These relations are valid only to first order, but that is sufficient to prove Einstein's point. The higher-order approximation proposed by SRT employs the Lorentz transformation (involving a Lorentz contraction of the train). That has no effect on the first-order conclusion. This completes the demonstration that sufficed to convince the world.

Case (b). Hertzian deductions

Recalling from Chapter 2, Section 2.5, Eq. (2.24), the Hertzian wave equation solution with its wave propagation speed $u = c + \left(\vec{k}/k\right) \cdot \vec{v}_d$ relative to S, we have for the right-going wave from the rear of the train $u = v_r = c + v_d$ and for the left-going wave from the front $u = v_f = c - v_d$. Here v_d is the train speed relative to S, and also the speed of the light detector, which is the eye of R, co-moving with the train. Inserting these values into Eq. (6.2a), we have, with Eq. (6.1)—on which all theories agree—

$$t_{rR} = \frac{L}{c + v_d - v_d} = \frac{L}{c} = t_1$$ (6.6a)

and

$$x_{rR} = v_d t_1.$$ (6.6b)

Similarly, from Eq. (6.3)

$$t_{fR} = \frac{L}{c - v_d + v_d} = \frac{L}{c} = t_1$$ (6.7a)

and

$$x_{fR} = v_d t_1.$$ (6.7b)

Consequently, if light propagation is described by the Hertzian field equations, we obtain

$$t_2 = t_3 = t_1$$ (6.8a)

and

$$x_2 = x_3 = v_d t_1. \tag{6.8b}$$

Thus S infers that simultaneity at first order is absolute and that the clock readings of the two inertial observers, S and R, agree as to the time (clock reading) at which the flashes are received by each of the two observers. (That is, S receives his flashes simultaneously at his clock time t_1, and R receives his flashes at both R's and S's clock time t_1. So t functions as an absolute or universal time parameter. This sort of time, as Newton said, " ... of itself, and from its own nature flows equably ... ")

At first order there is no time dilation, hence no basis for timekeeping asymmetry. Since there is thus complete symmetry between the two inertial observers, and all clocks run at the same rate, the roles of R and S can be interchanged without altering the conclusion that simultaneity is absolute and that both observers see the flashes simultaneously and at the same clock readings. This is the Hertzian deduction. However, within the context thus far developed it is only a first-order approximation that takes no account of any higher-order effect of motion on clock rates. In order to examine that, we turn next to the neo-Hertzian case.

Case (c). Neo-Hertzian deductions

Chapter 3, Section 3.3, Eq. (3.30), indicates that the neo-Hertzian wave propagation speed is $u = \sqrt{c^2 - v_d^2} + (\vec{k}/k) \cdot \vec{v}_d$. Consequently we have $v_r = \sqrt{c^2 - v_d^2} + v_d$ and $v_f = \sqrt{c^2 - v_d^2} - v_d$. Inserting these values into Eq. (6.2), we get, with the help of (3.3b), (3.4) and (6.1),

$$t_{rR} = \frac{L}{\sqrt{c^2 - v_d^2} + v_d - v_d} = \frac{L}{\sqrt{c^2 - v_d^2}} = \frac{L}{c} \frac{1}{\sqrt{1 - v_d^2/c^2}} = \gamma_d t_1 \tag{6.9a}$$

and

$$x_{rR} = v_d t_{rR} = v_d \gamma_d t_1. \tag{6.9b}$$

Similarly, from Eq. (6.3)

$$t_{fR} = \frac{L}{\sqrt{c^2 - v_d^2} - v_d + v_d} = \frac{L}{\sqrt{c^2 - v_d^2}} = \gamma_d t_1 \tag{6.10a}$$

and

$$x_{fR} = v_d t_{fR} = v_d \gamma_d t_1. \tag{6.10b}$$

So, in this case, we have $t_{rR} = t_{fR}$ (as well as $x_{rR} = x_{fR}$), indicating exact simultaneity of flash receptions at the eye of R, even at higher orders. From this simultaneity of flash receptions by R, S (using neo-Hertzian theory) infers the simultaneity of flash *emis-*

sion events in R's train system—hence the absoluteness of simultaneity of those events. But it is no longer true that the S-frame time $t_{rR} = t_{fR} = \gamma_d t_1$ of the event of simultaneous flash reception occurring at R's eye is equal to S's clock reading,

$$t_{rS} = t_{fS} = L/c = t_1, \tag{6.11}$$

for the time of occurrence of the *different* pair of simultaneous flash-reception events at S's own eye. (Note that there are two flash-emission events, but four flash-reception events, counting both observers, each using as his detector only one eye!) Since $\gamma_d > 1$ it follows that $t_{rR} = t_{fR} > t_1$.

We need to remind ourselves that the entire discussion in this Section has taken place from the viewpoint of S. The above evaluations of v_r and v_f apply only to the S-system in which one of the two light detectors (eye of R) moves with speed $v_d \neq 0$. So the inferences we have been making about what R sees are expressed in terms of S's clock time, $t_{rR} = t_{fR}$, and are not in general the same as what R actually "measures" with his own clocks, which run at a different rate. Rather than pursue the rather complex and sterile interplay of inferences that each makes about the other's measurements using naturally-running clocks, I shall defer further investigation in favor of a simpler treatment of this whole subject in the next chapter, based on introduction of a "collective time." That approach vastly simplifies such inferences.

As a review, it is of interest to compare all this with Einstein's first-order result. If we let $\beta = v_d/c$ and take $L = 1$, then Eqs. (6.4) and (6.5) yield

$$x_{rR} = \frac{\beta}{1-\beta}, \qquad x_{fR} = \frac{\beta}{1+\beta} \qquad \text{(Einstein)} \tag{6.12a}$$

for Einstein's analysis, and from Eq. (6.9) or (6.10) we obtain for the neo-Hertzian theory

$$x_{rR} = x_{fR} = \frac{\beta}{\sqrt{1-\beta^2}}. \qquad \text{(neo-Hertzian)} \tag{6.12b}$$

To recall: x_{rR} is the up-track distance at which train-rider R receives the flash from the rear of the train, and x_{fR} is the up-track distance at which R receives the flash from the front. Of course $x_{rR} > x_{fR}$ according to Einstein. Eq. (6.12a) gives Einstein's predicted x_{rR} and x_{fR} distances; whereas $x_{rR} = x_{fR}$, from Eq. (6.12b), is the neo-Hertzian result. The Hertzian prediction, (6.6b), (6.7b), is not shown ... it is just the straight line $x_{rR} = x_{fR} = \beta$, which is coincident with the other cases for small β. These results confirm what is immediately evident from (6.12), that the neo-Hertzian simultaneous flash receptions by the moving absorber occur at a time (or distance) that is the *geometric mean* of the times (or distances) of non-simultaneous flash receptions predicted by Einstein.

Summary. Using Maxwell's picture of light propagation as causally retarded at speed c, Einstein concluded from his train example that distant simultaneity of point events is "relative." We have shown, using identically the same example, that Hertzian (first-order invariant) electromagnetism predicts distant simultaneity to be absolute and predicts that clocks in different states of motion run at the same rate (to first order). Finally, neo-Hertzian (higher-order invariant) electromagnetism confirms the absoluteness of distant simultaneity, but assigns to clocks in different states of motion natural (proper-time) running rates differing by a γ-factor—a second-order correction. The moral is that the "relativity of simultaneity" is not an unassailable fact but the consequence of a particular picture of light propagation that is by no means the only one compatible with observation.

It may be added that variations of the formulation of Einstein's train example are possible and instructive. For example I proposed one variant[6.2] in which there is no necessary signalling by light or other means. Its outcome was as above, that distant simultaneity can be considered absolute.

6.3 Clock slowing: actual or symmetrical? (The twin paradox)

We need to face a foundational issue of relativism, *viz.*, whether to consider the clock-slowing associated with relative motion to be real (asymmetrical) or apparent (symmetrical). This is ultimately a matter of physics. Yet, where clock rates are concerned, special relativity theory (SRT) survives—*i.e.*, accommodates what is known of the physics—by being an amphibian between appearance and reality. Einstein's relativity principle, as customarily understood, demands a symmetry of asymmetry—the clocks of two inertial observers each symmetrically running slower than the other. To avoid an infinite logical regression to nonsense, SRT therefore needs clock rates to be *appearances*—lacking objective validity and contingent on viewpoint. Whereas, to earn credit for predicting the observed asymmetrical aging of the CERN muons[6.3] (circling and stationary in the laboratory), SRT needs clock rates to be *real* and objectively asymmetrical—which is to say, not contingent on viewpoint. For it is manifest that the clock of a muon circling at constant speed must run at a uniform rate ... and, since the net effect of the journey is *observed* to be a staying young, it must be that this staying young was a *process* taking place at a uniform rate throughout the uniformly circular motion. This can only mean that the SRT stay-at-home inertial observer's prediction of uniform rate-slowing of the space traveler's clock describes not an appearance but a reality—a factually real asymmetry. However, this conflicts with the "appearance" view of clock rates, dictated by the relativity principle. So, which shall it be? Symmetry or asymmetry? Appearance or reality? Agreement with principle or with observation?

SRT, of course, has a response—indeed, one for every shade and climate of authoritative opinion. Are we to believe, for instance (along with Feynman[6.4]), that the asymmetrical element in the muon or twin situation, *acceleration*, acts causally to convert appearance to reality? This parameter enters the theory nowhere explicitly ... it appears logically as a *deus ex machina*. Its only function is to save not the phenomenon but the theory. Pauli postulated *no explicit effect* of acceleration on timekeeping ... and the muon data confirm this for accelerations up to $10^{18} g$. Yet, here we are blandly alleging the most profound effect possible—conversion of appearance to reality, theoretical symmetry to factual asymmetry. That's a heavy load for a non-load-bearing member of any theoretical structure to bear.

Other apologists for SRT (following Wheeler-Taylor[6.5]) reject the acceleration mechanism entirely and cite an acceleration-free version of the twin paradox. They use SRT's event calculus (equivalent to a Minkowski spacetime diagram) to show that clock *phase jumps* properly account for the asymmetry of *elapsed time* observations. But they never explain how these discontinuous jumps fit with the necessarily uniform running rates of all clocks throughout the journey—implying *unobservability* by anybody of any clock rate discontinuities whatever. Neither actual clocks nor physical nor biological processes behave discontinuously in nature. The stay-at-home twin cannot reset his biological clock to accommodate the phase jumps needed to explain elapsed times. His clock must run *inconsistently*—slow to match the traveler's SRT analysis, fast to match the facts (or the stay-at-home's own SRT analysis). The relativist is plainly not concerned with describing all observable aspects of nature, only with describing whatever fits his calculus of point events (the Lorentz transformation). The latter enables hypothetical phase antics—knowable to analysts, unknowable to observers—to compensate rate inconsistencies, yielding correct elapsed times. (For a more thorough review of twin paradox "explanations" I recommend the delightful Appendix on the subject in Kelly's book.[2.13])

The Minkowski account is false to nature and true only to itself. It is false in two ways: it fails to describe what *is* observed (increased rather than decreased aging of the stay-at-home twin in the traveler's analysis—use of γ where $1/\gamma$ is needed), and it describes what is *not* observed (a clock phase jump or aging discontinuity). In effect relativists treat time *via* γ-symmetry where nature calls for $1/\gamma$-reciprocity. They are trapped inside the Minkowski box and have shown themselves incapable of thinking outside it. Whether this incapacity is constitutional or political, perhaps only a psychologist could say. Unfortunately, psychologists are no smarter than relativists. Consequently psychology has thus far proven of little help to hard science.

One more variant of the twin problem: Consider the CERN muon experiment,[6.3] but suppose that two bunches of muons circulate symmetrically in opposite directions. The laboratory inertial observer by symmetry sees each bunch as staying young equally, so that their aging is the same. This equal staying-young can be taken as observed fact, the symmetry argument being conclusive. According to SRT, each of two observers co-moving with these muon bunches sees the other as staying young. This means each is seeing the other as staying young relative to the factual staying-young seen by the lab observer, so each is staying younger than young and younger than the other, which is to say younger than herself. Thus the paradoxes of two observers are compounded into the contradictions of three observers. Experience is the great teacher. Experience has taught me that no logic, no silver bullet, no wooden stake will put the quietus on this undead monster. In contemplating such theory we ought to ask not simply how to apologize for it but how to apologize to future generations for our sluggishness in seeking an apology.

Such problems have in the past given rise to unhappiness among critics such as Herbert Dingle,[1.4] who have consequently been dismissed as demented by SRT's dedicated dogmatists. (Despite Dingle's having published a well-received book on SRT, which was cited by Einstein, it was decided after Dingle's defection that *he did not understand* SRT. Poor fellow, he just didn't get it. Consequently he was black-balled from his club of anointed English science sachems and Royal Society Pooh-Bahs ... a fitting punishment for failure to understand, as any modern card-carrying scientific clever-bones will agree.) SRT's all-pervasive symmetry ambiguity encourages its supporters to carry water on both shoulders: The twin-paradox saga shows clock-rate symmetry (during travel) and asymmetry (after return), and the CERN data[6.3] show asymmetry ... both, all, and anything else being triumphant confirmations of the theory, backed analytically by its event calculus and politically by the huzzahs of ten thousand physics professors and a hundred "science writers."

The epiphany needed to guide us out of this logical dilemma is a recognition that *clock rates* (or biological aging, *etc.*) by their nature fail to fit an event calculus of any kind. The very concept of clock rate implies not a point event nor interval-delimiting pair of events but a *process* sustained in time for an unspecified duration at an indefinite epoch. This process is not described by an event pair, only by an event sequence. SRT has no native aptitude to describe rates. Its rate follies are the direct result of stretching an event calculus past its elastic limit. Yet the theory's problem with clock rates is more poignant than a mere inaptitude. When winkled out of the point event mold and considered as physical observables in their own right, clock rates *are* described by SRT and are *described inconsistently*. That is the bitter pill supporters of the theory can never swallow ... It is the aspect critics cannot and should not forgive.

Since so much is made of the logical self-consistency of SRT as an event calculus (which is undeniable), it is worth pausing to pinpoint the *locus* of its inconsistency as physics. Specifically, consider the acceleration-free version of the twin paradox adduced by Wheeler and Taylor.[6.5] The traveler goes out, moving inertially and staying young. He passes the torch to another traveler of his identical age, who returns also inertially and similarly stays young during the journey, so that all observers agree on the youth of this second traveler compared to the stay-at-home. To be sure, the identity switch here is a bit of a swindle, since to eliminate it would eliminate the acceleration-free feature ... but let us not chop the logic of these authorities. To make matters clear to the meanest intellect, they offer a Minkowski diagram showing a large theoretical discrepancy between the two travelers' conceptions of "same time" at a distance. This discrepancy translates to a large clock phase (absolute aging) jump attributed by the travelers to the stay-at-home. Taking account of this phase jump allows the returning traveler to understand why his theory has predicted a very slow clock rate—meaning very little aging—for the stay-at-home, whereas the latter's observed aging is actually very great. The logical self consistency is perfect: SRT, applied by the travelers, predicts a very small *aging rate* of the stay-at-home, but it also predicts a very great *net aging*, the discrepancy being due to a clock phase jump requiring a "clock resetting" midstream or elsewhere (your call). The logic, and the nonsense, of jumping clocks, are both perfect ... of that degree of perfection one looks for in the most refined works of idiot savants.

We can imagine a dialog of the following sort between the outgoing traveler T1 and the incoming traveler T2:

> T1: Hail, fellow traveler!

> T2: Well met, brother. What can you tell me about that bibulous fellow you left on Earth, whom I journey to visit?

> T1: The sot has drunk from a veritable Fountain of Youth and aged very little during my journey—and will age equally little during your remaining journey. However, when I pass on to you this Torch of Faith, he will get his come-uppance. That act will cause his days suddenly to be numbered. By instant action at a distance his goatish gambollings among the lambs of spring will abruptly cease and the frosts of age will blight his few remaining palsied hours.

> T2: Should we really do this to him? It seems a dirty trick.

> T1: True, but have no fear ... he is a low fellow of little imagination who will suspect nothing and have no basis for recrimination. To him it will seem that he has lived a life as full as that of any other sinner. He will be under the illusion of experiencing uniform clock rates throughout his lifetime.

> T2: Wondrous indeed is thy name, O Science.

> T1: Yea, verily, more so than that of Science Fiction. Such revelations are the emoluments of those who sustain their monastic faith in theory, unbroken by the harsh impingements of worldly experience.

T2: True ... yet flesh is weak. My shining faith is sometimes tarnished by doubt. Why is it that this act of our passing the Torch will produce such drastic distant consequences?

T1: Reason it out, little brother. Pure reason is the Kantian soul of Science: Human agings are by definition continuous except at events of discontinuity. Our motions, yours and mine, are continuous except at our present event of passage. What else may be logically concluded than that this unique event of discontinuity bears a causal relationship to our partner's aging discontinuity, which we know to be factual because the noble theory predicts it?

T2: Blessings upon you, Font of Wisdom, for reaffirming my faith through the implacable power of reason. May yours be the timeless serenity of the closed mind, may you be rewarded with instant tenure and a full professorship at that great Harvard-in-the-Sky to which you journey, and may your endless worldline prosper eternally under the Diagram of Minkowski: *In hoc signo vinces.*

T1: In the name of Einstein, the Father, Minkowski, the Son, and the Holy Spirit of Spacetime, farewell. May we meet again on the far side of Riemannian curved space ... meanwhile, join me in prayer that nobody pulls the same dirty trick on us.

T2: Amen, brother, amen. *Pax vobiscum.* And look out for wormhooooooles!

SRT's logical self consistency is never in question. But *self* consistency is not enough for physics. A theory's "self" is a small, vain thing. It swells and takes on greater proportions only to the extent that it does not conflict with the general background of human experience. In this case we happen to know about biological aging processes that they do not suffer discontinuities. The travelers know this, so they have every human right to be dissatisfied with a theory that tells them the stay-at-home continually aged at a very slow rate during every moment of each of their journeys, yet was found in the final event to have aged mightily. For experience tells them about the aging process that it takes place uniformly and continuously in nature—so if it was factually slow throughout the journey it is inconsistent (not with logic but with experience) to assert a discontinuously large factual net aging.

The traveling inertial observers cannot use SRT to get predictions in agreement with their broader knowledge of the continuity of biological aging processes. By no means can the theory's event calculus be induced to reveal to either of the travelers a *number* describing that *actual uniform rate* at which the stay-at-home's clock runs as a matter of physics—the number that quantifies the stay-at-home's "illusion" of a uniform aging rate. That "illusion" is a factual ("measured") aspect of nature in which the theory shows neither interest nor capability. Instead of revealing to the traveler the facts about the stay-at-home's clock rate, SRT flaunts another (smaller) "rate" number in direct conflict with what is observed by anyone. That is, it disports itself incestuously with events, getting its story in that department straight by fantasizing about clock phase jumps. Although the theory tells the travelers the truth about total elapsed times, it lies to them about clock rates—it fails to tell them what was factually observable and tells them instead something *inconsistent with experience* and consistent *only with the theory*. Telling the truth about Peter thus excuses lying about Paul.

Inconsistency with experience in no way discourages a relativist. We have already remarked that SRT denigrates experience. It has to do so in order to survive. The most striking example is its allegation of the invariance, hence the physical "reality," of the spacelike interval $d\sigma$. Practically by definition there is, and can be, no experience to back that up. In such a case experience is discounted—it does not matter. This style of approach is pure metaphysics, in the pejorative sense. Even relativists cannot stick by such an attitude, for when challenged to show evidence in support of their theory they have nothing to fall back on but experience. So they play fast and loose with experience, appealing to it when convenient, ignoring it when convenient. Experience is their *bagatelle*, their political football.

But a day of reckoning is at hand. Although physicists can keep this up, generation after generation, never breaking stride nor blinking an eye, there are also engineers in the world, and they are answerable to the world as it is. Some of them have designed a Global Positioning System (GPS) that uses traveling clocks. For anyone who respects experience, experience with the GPS can have a decisive bearing on the fundamental issues of relativism. Empirical data from multiple sources are now available showing timekeeping asymmetry to be an objective physical fact. So, it is high time to end the symmetry-asymmetry ambiguity. In the next section I call attention to direct evidence provided by the GPS.

6.4 GPS evidence for clock rate asymmetry

The Global Positioning System (GPS) has been fully operational now for over a decade and has yielded data directly relevant to the issue under discussion. Atomic clocks in various earth satellites have work done on them (in the course of placing them in earth orbit), which alters their gravitational potential energies and states of motion. In presumed consequence of such alterations their running rates are changed in accordance with known laws. The effect of "time dilation" resulting from motion (atomic clock rate slowing by a relativistic γ-factor), mentioned above, is verified, and also the effect of raising the clock to a position of weaker gravity (clock rate increase) or of immersing it deeper in a gravity field (rate decrease). Since clocks in various earth satellites must all be able to "talk to each other" and to master clocks on the ground, it is highly advantageous as a practical matter to have them all running continually at objectively the same rate. Nowhere does SRT hint at such a thing as either possible or desirable; but fortunately, like Dingle, GPS engineers do not "understand" SRT. As discussed above, it is their practice to modify the running rates of the satellite clocks, *while still on the ground*, by "tuning" them to new permanent running rates chosen to compensate the anticipated effects of change of both gravity and state of motion. This is done routinely, and the results—a working system—speak for themselves.

Let us contemplate what this means. When the to-be-orbited clocks, while on earth, are altered in their running rates, they thenceforth permanently cease to be Einstein clocks. An attempt to use such a clock to measure the speed of light on earth would not yield the "universal" value c, but some other value. (Speed of any kind is a quotient of length and time. In a given stationary apparatus for measuring the speed of light no length change occurs, but a clock rate alteration would change the numerical quotient.) Already we are well beyond the SRT pale. Such altered clocks defy established canons of thought: they are not "clocks," because they are not Einstein clocks. Their existence is forbidden by Einstein's second postulate of light-speed constancy. The time they tell is not "time," but something for which SRT's acolytes lack even a name.

Next, this spayed (non-Einstein) clockette—dare I call it a clock?—is placed in orbit and it runs as intended, at the same rate as earth-surface clocks (and at the same rate as the other compensated satellite clocks, as all can mutually confirm by radio signals). This is an objective fact, validated by experience. The system would not work otherwise. It means that the clock-rate alteration due to motion, quantified by the γ-factor, is also an objective fact, because it was objectively compensated and its effects thereby caused objectively to go away. Uncompensated or proper-time clock slowing due to motion is thus not an appearance but a verified fact.

In further discussion here we shall set aside for now the gravity effect on timekeeping, supposing variations of location in a gravity field to be separately compensated. (This is legitimate because both gravity and motion compensations produce small and independent effects, which superpose linearly.) The compensation for motion, then, acts to speed-up the orbiting clock's rate so that it keeps up with the faster-going earth clock, thus countering the natural tendency of the orbiting clock to go slow due to the motional "time dilation" effect. Since the compensatory speeding-up was an objective fact, and its resulting equalization of all clock rates was an objective fact, it is apparent that for a true Einstein proper-time clock (*i.e.*, any uncompensated clock) its slowing in orbit is also an objective fact. And at no time does the earth-surface proper-time clock alter its fast running rate. Despite the relative motion, it does not run slow nor "look like" it is running slow to any of the moving satellites—nor is it "measured" by them as running slow!

Hence it is in direct disagreement with fact to assert reciprocity, as SRT does, such that *symmetry* prevails between the earth clock which "sees" the satellite clock slowed and the satellite clock which "sees" the earth clock slowed. The satellite clock must see and measure the earth clock not as slowed but as continuing to run fast—for the simple reason that the (compensated) satellite clock has been deliberately speeded up to match; and it can verify, while in orbit, that this speeding up has brought it into *rate equality* with the ground clocks. To repeat, there is an objectively real *asymmetry*—otherwise the compensation would not work. If the satellite clock (compensated or not) were to "see" the earth clock as slowed, there is no way the orbiting compensated clock could "see" all (earth and other orbiting compensated) clocks as running at the same rate as itself.

Despite my many and repetitious verbalisms, some readers may be so deeply indoctrinated with SRT as not to find the above claim of a genuine clock rate asymmetry convincing. If this is the case, let me offer a couple of clinchers to the argument, at the risk of wearing out the patience of any reader already convinced.

Timekeeping Asymmetry Clincher #1. The most direct way to see the relativity-violating asymmetry is to recognize that the earth-surface proper-time clock X must be running objectively faster (not appearing to run slower, as relativistic symmetry would have it) than the orbiting proper-time clock Y (X > Y), because the orbiting compensated clock Z is observed to run at the same rate as the earth-surface proper-time clock X, keeping continually in step with it (X = Z), and Z's compensation was such as to make it run faster than the orbiting proper-time clock Y (Z > Y). (This was true on earth where the compensation operation was performed, and remains true when compensated and uncompensated clocks are orbited together.) Thus if X has the same rate as Z, and Z runs faster than Y, then X must run objectively faster than Y. (In symbols, X = Z > Y implies X > Y.) However, X cannot *objectively* run at a different rate from Y (X ≠ Y), *without* violating the customary understanding of the relativity principle—to wit, (a) that proper-time intervals are invariant, hence natural clock rates are the same in all inertial systems, which are "equivalent" as to their physics; (b) that asymmetries among inertial systems are only appearances; (c) that there is hence an underlying symmetry of (apparent or "measured") asymmetries.

But, I forget, the foregoing is an application of Aristotelian logic, which has long gone the way of Euclidean geometry. Both are now relegated to footnotes to the higher logics. I have no doubt the latter can readily cope with the above criticism or any others. The result will be the same in all cases: to reveal that the SRT critic has no understanding of the theory. Indeed, by definition no adverse critic could possibly understand a theory that is not a theory but a fact. Facts can be misunderstood but never criticized. Historically, Aristotelian logic served to establish SRT, *via* Einstein's train example. But when it fails to keep it established, the mobilization of more powerful logical means becomes justified to that good end.

Preparatory to a second clincher, it will be useful to review in some detail how GPS clocks are compensated. We restrict attention to atomic clocks (specifically, cesium clocks) as our basic measuring instruments. The notional procedure is as follows: First a reference environment is chosen—say, a state of rest on the earth's surface at a certain place (altitude, gravitational potential) and any clock placed permanently in that environment and using a chosen atom (cesium) is elected to be a Master Clock. (A Master Clock fixed on the earth's surface will itself in general need compensations for earth spin, *etc.*—but I omit this complication here.) The reference proper-time "second" of time is then defined as some agreed number N_0 of natural hyperfine resonant oscillations of the cesium-133 atoms of the Master Clock in its reference environment (N_0 being some large integer). Thus, whenever N_0 oscillations have occurred, a *counter* contained in the clock is set to mark the time-passage of one "second."

In what follows I shall continue to leave gravity out of the discussion. (That topic will be addressed in Chapter 8.) In order consistently to quantify timekeeping in a different motional environment characterized by squared-velocity v^2 motion relative to the Master Clock, and by an associated time dilation factor $\gamma = 1/\sqrt{1-(v/c)^2}$, a compensation is conveniently introduced prior to transfer, while the secondary clock to be transferred to the new environment is still co-moving with the Master Clock. This compensation consists in setting the secondary clock's counter to redefine the "second" as $N_0' = N_0/\gamma$ oscillations of the cesium atom. That is, in agreement with Eq. (3.3b), the factor f by which the running rate of a clock, altered by having mechanical work done on it, is compensated to restore the original rate, is

$$f = \frac{d\tau}{dt} = \frac{1}{\gamma}, \qquad N_0' = fN_0. \tag{6.13}$$

The *second* thus redefined—once the clock is placed in its intended new γ-state of orbital motion—will measure the earth-surface second of "time" and hence will keep the orbiting compensated clock permanently in rate synchrony with the Master Clock. (This rate synchrony does *not* refer to the basic atomic oscillations, which are motion-affected, but to the clock "ticks" occurring at "one-second" intervals.) Given a known further change in state of motion, N_0' must be assigned a different value.

Why does this compensation procedure work? Simply because the relative motion of the clock, when it has been placed in its new state of motion by doing work on it, causes its internal oscillator to run *objectively* slow by the γ-factor. (The relative motion between orbital clock and Master Clock is not *postulated*, as in Einstein's kinematics, it is *produced*, by work being done asymmetrically, on the orbited clock and not on the Master.) That is, the cesium atoms in the clock on which work has been done will, in the perverse nature of things (the faster a clock "travels," the slower it "runs"), become lazy and oscillate more slowly by the γ-factor ... so, in order to cause this work-tranquilized clock to simulate the timekeeping of the faster earth clock, we have to fool it as to what the "second" is. Since we cannot speed up the actual atomic oscillations—that being against nature—our only recourse is to count the "second" as a different (smaller) number of oscillations. Because the time-dilated oscillations have longer periods, fewer of them are needed to quantify a given stretch of earth-surface time ... so we divide N_0 by $\gamma > 1$ to effect our compensation. As noted, this is merely a matter of resetting the clock's internal counter (which tallies an integral number of oscillation events—the integer closest to N_0/γ—that defines what we choose to mean by the orbital clock's "second").

From the above we recognize the feasibility of fabricating a single clock that performs two separate and independent timekeeping functions: It contains a single cloud of cesium atoms but uses two counters, one set to count the second as N_0 oscillations of the hyperfine resonance (thus measuring proper time τ), the other set to count the second as $N_0' = fN_0 = N_0/\gamma$ oscillations, thus measuring rate-compensated time t_0, to which we shall give the name "collective time," to suggest its shared nature among differently-moving clocks. Here, as above, $\gamma(v^2)$ describes the effect on timekeeping produced by a relative motion v^2. We may call this a *dual-function clock*. Chapter 7 will elaborate on this concept.

Timekeeping Asymmetry Clincher #2. One form of the relativity principle may be viewed as asserting a symmetry of *operational procedures* between any two inertial systems, such that similar operations produce similar results. Thus the (notional) orbiting observer and the earth-surface observer should be able to do similar clock compensation operations and get similar measurement results. Let us look into this. Suppose we have on earth side-by-side two dual-function cesium clocks of the sort just described, initially at rest on the top of a tall mountain. One we leave at rest, so that it continues to run at its normal rate, showing both earth-surface proper time and γ-compensated (rate-speeded) collective time. The other dual-function clock we place in low orbit—at the same gravity potential, to eliminate effects of gravity difference from the problem. Because of time dilation—a physical phenomenon causing a reduction of the natural oscillation rate of the cesium atoms—both functions of the dual-function clock have their atomic oscillations slowed by a γ-factor when the clock is placed in orbit. (The two "clocks" of a dual-function clock are equally affected, their rates being proportional, because both derive their basic running rates from the slowed oscillations of the same cesium atom cloud.) This rate reduction of the orbiting clock means that its proper-time clock runs slower than the earth-surface proper-time clock by a γ-factor, and that its compensated clock runs (counts seconds) at the same rate as the earth-surface proper time clock (per GPS evidence). That is what the compensation was designed to achieve.

Next, consider the orbiting clock to be returned (by rocketry) to its original state of motion, side-by-side with the earth-surface clock. We may suppose that this restoration of the original environment of the cesium atom cloud restores its original oscillation (running) rate. Alternatively, suppose a dual-function clock to be built *ab initio* in orbit, by the same operational procedures used for the corresponding clock-building on earth; and let it be transported to earth's surface. In either case, after return to earth, the "traveling" dual-function clock's proper-time rate will match the proper-time rate of the stay-at-home clock, and its γ-compensated clock rate will match the γ-compensated rate of the compensated stay-at-home clock. This agrees with the twin paradox assumption that the space traveler *after returning home* will age once more at the same rate as the non-traveler. It implies a speeding-up of clocks transported from orbit to earth; to be contrasted with the slowing-down that occurs as a result of transport from earth to orbit. (We shall have more to say presently about the "reversible work" involved in such transport.) Thus the two types of transport produce asymmetrical effects on timekeeping. This factual reciprocity, $\gamma \rightarrow 1/\gamma$, already conflicts with accepted understanding of the relativity principle, which asserts a measurement symmetry, $\gamma \rightarrow \gamma$.

Let us specify certain clock rates for easy reference:

A = rate, when transported from orbit to earth's surface, of a clock built in orbit and compensated there by increasing its rate by a factor of γ.

B = rate, when not transported, of an uncompensated clock built in orbit and showing natural or proper time there.

C = rate, when not transported, of a clock built on earth's surface and compensated there by increasing its rate by a factor of γ.

D = rate, when transported from earth's surface to orbit, of a clock built on earth's surface and compensated there by increasing its rate by a factor of γ.

E = rate, when not transported, of an uncompensated clock built on earth's surface showing natural or proper time there.

Suppose, as before, that the orbiting observer mimics the other and builds a dual-function cesium clock in orbit, set to read both proper time and (rate-speeded) compensated time. We continue to exclude gravity from the problem or suppose it separately compensated. If relativistic symmetry prevailed, the orbiting compensated clock, when retro-fired in a rocket and returned to earth's surface (clock A), would run in step with the orbiting proper-time clock B (A = B). [That is, by the relativity principle, the symmetry D = E demands a matching symmetry A = B.] But in fact clock A must run in step with the permanently-resident earth-surface compensated clock C (A = C), because the compensation operation (substitution of N_0' for N_0) was numerically identical in both inertial systems, and all clocks sharing a given state of motion run at the same rate when set in the same way. [Here we assume that any naturally-running clock, wherever built and however orbited, when returned to earth's surface *via* any history of motion, without fiddling its rate, will run *after return to earth* at the same rate as the native proper-time clocks permanently resident there. On this we simply follow Einstein, who founded his SRT on an implicit "state function" assumption—one of several unstated assumptions—that natural clock rates are a function of their state of motion. This means that all uncompensated clocks, regardless of history, when placed in a given state of motion, run at the same (natural or proper-time) rate, other things such as gravity being equal. This state function attribute must apply also to all clocks set to run in the same way—*e.g.*, compensated by the same γ-factor.]

Since the compensated-in-orbit-and-transported-to-earth clock A runs in step with the earth-surface compensated clock C, A = C, it cannot run in step with the orbiting proper-time clock B (as demanded by relativistic symmetry), because the latter by its setting runs slower than the orbiting compensated clock D (B < D). The clock D is known from GPS data to run at the same rate as the earth-surface proper-time clock, E (E = D), which in turn by its setting runs slower than the earth-surface compensated clock (E < C). Thus our inequalities yield B < D = E < C = A, or B < A, which contradicts the symmetry requirement B = A of the relativity principle. In short, operational symmetry fails, and with it Einstein's form of the relativity principle. Incidentally, our argument shows that B < E, hence B ≠ E, so proper-time symmetry fails, as we have repeatedly claimed.

The operational procedures of clock compensation that work for clock transfers from earth to orbit do not work for identical transfers in the other direction. SRT's alleged symmetry of inertial systems in respect to clock running rates, or in respect to operational procedures of clock setting, is not there in nature. The fault traces to the accepted form of the relativity principle itself, or to its misconceived application by SRT. Hence that principle, which dictated the spurious alleged symmetry, must be reformulated. (This will be one of our subjects in Chapter 7.) In effect SRT is directly refuted by GPS evidence that has been in the public domain since 1993 ... an interval during which the physicists' universal hype claiming total confirmation of SRT built to a crescendo in 2005, the *annus mirabilis* + 100.

In summary, the SRT tale of "seeing" or "measuring" timekeeping symmetry is a myth. Manifestly, there is a real, measurable asymmetry of physical clock rates. It must be accepted as a fact that when work is done on any clock (compensated or uncompensated) to alter its state of motion, it runs thereafter objectively at an altered rate in the new motional state. (We shall presently conclude that a comparable thing is true in regard to gravity potential energy changes.) An asymmetry of physical energy states produces an asymmetry of clock rates; *i.e.*, motions leading to different total energy states produce different clock running rates. This seems to be the experiential-world reflection of the underlying physics

This has been a rather taxing section of my book, both to read and to write. I hope the reader has been able to follow it, for it contains the crux of what needs to be communicated in identifying one of the most basic and enduring flaws in current theoretical physics. I think nobody who takes the trouble to study it can miss the point.

6.5 Clock rates, free-falling *vs.* supported in a gravity field

Let us probe a bit deeper. The relativity principle concerns purely inertial motions, and it would seem that not all GPS clocks satisfy that. It is likely that SRT's defenders, quick to detect loopholes, will feel that in the example just given I have wrongly attributed symmetry to "inertial systems" by ignoring the distinction between a true Einstein inertial (free-falling) proper-time clock in orbit and a proper-time clock supported at rest in a gravity field, *e.g.,* our hypothetical surface-clock supported on a mountain peak. Actually, as we next show, for an arbitrary state of motion along a gravity equipotential surface, a clock's natural running rate does not depend on its degree of "horizontal" support (transverse to that surface) in the gravity field. First, two theorems, one false and one true:

Theorem (False). The proper-time running rate of a clock, dependent on its potential energy (depth or position) in a gravity field, when in free fall, does not depend on its state of motion.

"Proof:" A clock in free fall can be spoken of as at rest in a locally inertial system. As a result of its inertiality, this free-falling clock, if subject to no deliberate "corrections" and allowed to run at its natural rate, tells proper time by definition. Such a clock, viewed as being at rest in its local inertial system, cannot have its rate affected by its absolute state of motion, *according to the usual understanding of Einstein's relativity principle,* which asserts all inertial systems to be equivalent for all measurement purposes, *including timekeeping.* So, the proper-time running rate of the free-falling clock is independent of its state of motion, *q.e.d.*

Commentary: I am obliged to remark about this "proof" that it relies on the relativity principle, as normally interpreted, for its claim that the proper-time running rate of a free-falling clock does not depend on its state of motion. Actually, this is directly and definitively contradicted by the GPS evidence, since the γ-

factor correction used in pre-compensating the clock *does* depend on the clock's state of (anticipated) orbital motion. The compensations are different, for instance, at a given satellite's aphelion and perihelion, where the speeds differ (as does gravity, but we continue to view that as separately compensated). Both gravity potential change and state of motion change affect the running rate of the GPS proper-time clock when placed in orbit. Different orbits require different compensation factors; yet if proper-time clock rates were the same in all orbits, as the theorem asserts, the compensation factors (needed to bring them all into step with the earth-surface Master Clock) would be *the same*. For this reason, on empirical evidence, I consider this "Theorem" to be physically false (although true according to the premises and logic of SRT, as normally understood).

Theorem (True). The proper-time running rate of a clock, dependent on its potential energy (depth or position) in a gravity field, does not depend on its degree of "support" at a given value of gravitational potential (*i.e.*, does not vary from free fall to full support at a given "height" in a gravity field, regardless of its state of transverse motion on an equipotential surface).

Proof: Consider our clock supported in a gravity field on a mountain top. (The "mountain top" may be traveling, with or relative to the earth, at any transverse—horizontal—speed.) It runs at some natural rate. Let it be dropped under gravity—that is, put instantaneously into a free-falling local inertial system. Initially, this new rest system of the clock is at the same gravity potential as, and at zero relative speed with respect to, its previous state of motion. Therefore initially the $(\gamma - 1)$-factor describing its slowing due to change of motion, and any effect of gravity change, must both vanish. So neither motion nor gravity change can have an instantaneously observable effect on the clock's running rate. This is to say, immediately after the clock is dropped, its running rate remains essentially invariant. But we know that when dropped it *instantly* becomes inertial and runs by definition at its proper-time rate for that position in the gravity field. Whether dropped or supported, inertial or non-inertial, all naturally-running clocks at the same place in the field run at the same rate. Hence, our result is established *without appeal to any relativity or equivalence principle*: degree of "support" in the gravity field has no effect on clock proper-time running rate, *q.e.d.*

In consequence of this theorem the fact that some clocks of the GPS earth-satellite complex are truly inertial and some are supported in a gravity field makes no difference from the case in which all clocks are truly inertial. For the latter case Einstein's form of the relativity principle is clearly violated by what is observed. In other words, if we drop the earth Master Clock, then instantly all clocks of the GPS become inertial, and the uncompensated Master and compensated orbiting clocks stay in step, contrary to Einstein's form of the relativity principle, which would require the dropped clock's proper-time rate not to stay the same but to decrease instantly, upon becoming truly inertial, to match the decreased proper-time rates of the other truly inertial (orbiting) uncompensated proper-time clocks—as asserted by our false theorem.

Further commentary: SRT is a hard theory to refute. One has the sensation of shadow-boxing inside a fogbank. As mathematics, SRT is simple and well-defined; but as physics it takes on all shapes from Máh to Máhi, all shapes of Vishnu. It is the quicksilver of physical theories ... wherever you stomp on it, it is someplace else. Suppose, for example, that after much expository labor (as above) you have convinced one contingent of its supporters that timekeeping asymmetry among inertial systems is not an "appearance" but an objective reality. Then another contingent is sure to come charging up with,

> Stop wasting our time. We knew that all along. Of course time dilation is objectively real, and of course the Lorentz contraction is objectively real—that's exactly what the equations of the theory are saying. It's trivial: These two real physical phenomena work together to ensure that the quotient of distance and time covered by propagating light is a universal constant c. Why all the fuss? You obviously don't understand the most elementary aspect of the theory.

It is worth remarking, in commentary on the position just quoted, that a belief in the objective reality, hence asymmetry, of time dilation entails a matching belief in the objective asymmetry of the Lorentz contraction—unless the believer is prepared to abandon *velocity reciprocity* among inertial systems, and with it the Lorentz transformations. This particular contingent is on thin ice in respect to the facts of empiricism (see Section 6.7, where mention is made of the failure to date of attempts to verify the Lorentz contraction as an objectively real physical phenomenon).

They are on the same sort of thin ice as the general relativists who stake everything on the next round of refinement of gravity wave detection—ever retreating into the next decimal place. The main difference is that the general relativists are at least seeking empirical confirmation, whereas the special relativists nowadays rest on their oars in that department and rely on myths about experiments instead of experiments. Does living memory go back to the last time an SRT supporter demanded empirical evidence for the Lorentz contraction? The SRT schools of thought are legion and their obligation to mutual consistency is nil. The schisms among them mirror those characteristic of the more honestly-acknowledged religions.

6.6 Platonic time and simultaneity

A new view fostered by the GPS evidence is that actual clocks are related to a Platonic ideal conception of "time" (our collective time t_0) in much the same way actual thermometers are related to an ideal conception of "temperature." In each case compensations are necessary and possible, owing to knowledge of the applicable physical laws, to enable the actual to approach the ideal. If we want to describe either time or temperature in the simplest way, it behooves us to make the corrections without apology. In the case of time, we already know for balance-wheel clocks to make corrections for friction, *etc.*, without fussing. But for some reason we have balked at corrections for changes of state of motion, and have developed a mythology of "symmetry" of time-keeping among inertial motions for use on all occasions not involving comparison with experiment.

From the present standpoint, Einstein's error lay in needlessly over-complicating physical description by failing to recognize the desirability of compensating for "apparatus effects" of motion and gravity on timekeeping—effects whose existence was only beginning to emerge in 1905. Consequently motional effects, instead of being eliminated, became integrally incorporated into time's definition. The resulting non-optimal definition of "time," abetted by a non-invariant electromagnetism, produced a complex, contorted, ugly physics that was mathematically nevertheless beautifully self-consistent ... and for that reason was inordinately admired for its beauty. Let us pause to analyze: SRT's event calculus is designed to tell us, when we know four coordinate numbers describing the space-time location of a point event in inertial system K, what the corresponding four numbers are that describe the same event in inertial system K'. It will do this infallibly, with total consistency with the theory's axioms and without contradiction. That is what it will do and all it will do. Ask it for more and it defaults. Ask it to describe clock rates and it defaults ... rather, it abandons consistency. It took a century before the actual falsity-to-nature of the resulting theory clearly emerged. Indeed, that process of discovery is still going on—the onset of its very beginnings not yet (at this writing) having dawned on the professional physics community. SRT deserves its place in history as a testament to the persistence of mutually compensating errors, abetted by an amazing forgiveness and obliviousness of error by its "scientific" patrons.

Lest I be accused of uncritical Platonism here, I pause to insist that my objection to Einstein's treatment of inertial systems (as embodying the perfect symmetry permitted by ignoring the path-dependent machinery of motional history) is precisely that it is *too* Platonic. That is, it is too dissevered from operational reality. One never encounters in nature anything like Einstein's free-floating *Gedanken* inertial systems, nor is one's state of ignorance ever so profound as to warrant the hypothesis of *no conceivable knowledge* of the physics of actions by which objects (including "reference frames") acquire altered states of motion. Nor, for that matter, can such free-floating systems be reconciled with Einstein's own recognition of the path dependence of proper time. *Relative motion change* is more likely to be produced by perspiration than by postulation. To establish differences in state of motion, work always has to be done and energy state changes have to occur—be they only *Gedanken* changes. That such energy changes of "frames" can be ignored, as Einstein and the mathematicians have it, is an hypothesis that is contrary to what is known or readily inferred on the basis of today's knowledge of the empirical facts of timekeeping.

My philosophical position, then, is one of a moderate advocacy of tolerance for a bit of Platonism and a bit of operationalism or instrumentalism—always with insistence on certified reliability of the instruments. It was the uncritical identification of idealized "inertial systems," divorced from how objects at rest in them attained their different states of motion, that contributed heavily to the prosperity of Einstein's basically indefensible SRT. One has to take one's Platonism eclectically and examine one's ideals case by case—critically. If we would be physicists, we need to keep our minds open, stay philosophically loose, and be ready to jump with the cat. The readiness is all.

If we elect to stop playing the SRT game—to make new rules and play a different game—then Einstein (proper time) clocks become simply uncompensated clocks that *fail* to tell properly the simplest and most useful kind of "time" ... much as uncompensated thermometers—though they undeniably tell temperature—fail to tell properly the most useful kind of "temperature." The pernicious consequences of this time-telling failure include loss of a consistent universal "now"—the *now* of our everyday personal experience—the discrediting of perception and the taking on of an elaborate load of ideological baggage, of gripping interest to philosophers, scholars, teachers, mathematicians, journalists, popularizers of science, and a motley horde of other camp followers, but of no use to physicists.

Once we recognize that Einstein's timekeeping (clock rate) symmetry between inertial systems is not there, and accept the idea of collective time (elaborated in the next chapter), we become able to recognize a more important symmetry between inertial systems that *is* there—*viz.*, the symmetry of *distant simultaneity*. This is the physically significant symmetry, vital to the existence and reciprocity of "now" and to an understanding of action-reaction balance, quantum non-locality, gravity's instant force action, and even the definition of causality/acausality. Without its acceptance, quantum mechanics has to be looked upon as some sort of temporary aberration—a passing fad. Einstein banished distant simultaneity to limbo, and only a virtually unprecedented resurgence of mass sanity can reinstate it.

6.7 Length Invariance

The compatibility of distant simultaneity, one of our key topics in this chapter, with length invariance—indeed, the direct equivalence of the two concepts—may be seen from a completely elementary consideration: Let two equal meter sticks slide past each other at any speed less than c. If their lengths are invariant under this relative motion, the events of coincidence of their end points will occur simultaneously for observers riding with each stick—indeed, for all observers. It is only through Lorentz contraction, or the equivalent, that any other conclusion can be reached. Conversely, regardless of how relative motion affects clock running rates, if proper-time readings of clocks at the ends of each stick, pre-synchronized (in the usual way of Newton or Einstein) independently by each of the two stick-riding observers, are equal for the coincidence events (not necessarily the same numerical readings for the two observers), this can happen only if the sticks are of equal length.

From this it would appear that length invariance, a postulated input to our neo-Hertzian theorizing of Chapter 3, logically entails the absoluteness of distant simultaneity. I have not, however, striven in this exposition to axiomatize my propositions nor to pare down my arguments to minimum logical essentials. Newton the Lion could take that path, because he had no seriously established opposition (Descartes and such aside) ... my task is more difficult. It is less a job of physics and more one of sociology or politics, an exercise in the art of uphill persuasion. Few people find axioms persuasive. Formal axiomatization is for the persuaded. Nevertheless, there are certain jobs that only axioms can do.

Up to this point I have baldly postulated length invariance, in the teeth of SRT belief and without effort at justification beyond pointing out, as above, that distant simultaneity implies length invariance and *vice versa*. Let me pause here to try to give this a deeper shade of plausibility before going on. When two distinct point objects, initially at rest, are subjected to identical accelerational histories parallel to a line joining them, their separation distance does not change, as measured in the original rest system. This is what common sense, Euclid, Galileo, and Newton all say, and for once Einstein agrees. That is, according to SRT, the initial rest-system observer will neither "see" nor "measure" any alteration of the distance between the two point objects. Lorentz contraction is there none. So, one kind of "length" transforms invariantly. Another kind of length that so transforms is length transverse to relative motion. So here, in the midst of SRT, are two distinct kinds of length that behave invariantly. Why not all kinds?

Why not, indeed ... The reason is the physicists' commitment to certain mathematics—the Lorentz transformations. The latter started out describing a physical causal effect of the ether in longitudinally squeezing-together any body that tried to move too fast with respect to the fundamental rest system of that ether

(Lorentz-Fitzgerald contraction). There are still believers in a Lorentz ether possessed of a definite state of motion. (I am not one, as I prefer not to multiply entities without necessity.) Einstein kept the math of this but ditched the physics, so the Lorentz contraction ended up with no cause whatever. Bereft alike of physical cause and empirical support, the Lorentz contraction marches bravely on today, sustained by a dogma—that of the "metric nature of spacetime." For those who abandoned hope of extracting a causal explanation from SRT, the Lorentz contraction became a "kinematic effect"—which is a scholarly-obscurantist way of saying it just *is*.

And I say it just *isn't*. When you trace the Lorentz transformations back to their point of origin, it is in the disabilities of Maxwell's equations—in the under-parameterization that renders them spuriously spacetime symmetrical. Remove that deficiency of electromagnetic description by improving the adequacy of parameterization and you remove all basis for Lorentz covariance and substitute the nobler aim of *invariance of all equations qualified to describe nature (i.e., objective reality)*. Once the conceptual obstacles are thus removed—and since we know (*i.e.*, all parties acknowledge) that in certain circumstances there are certain kinds of length that transform invariantly—I say let us suppose the simple, the economical, the elegant thing: *All length transforms invariantly in all circumstances*. I predict that that will do very nicely for a starting description of nature. Gimme that old time religion—it was good enough for Euclid and it's good enough for me. If in the far distant future this proves demonstrably, empirically wrong, I solemnly pledge to come nonlinearly curving back out of the fourth dimension, pop through a wormhole, and eat the hats of Gauss, Riemann, and John Archibald Wheeler.

I suppose there exists a stubborn rear guard dedicated to the proposition that the Lorentz contraction is proven empirically by the Michelson-Morley (MM) experiment, mentioned in Chapter 3, Section 1. In this area of argumentation, MM is the last resort of a scoundrel. One can cut through reams of debate on this topic by considering what the experimental apparatus actually did. It was an interferometer with orthogonal arms and it sat in a pool of mercury. That's what it did. It sat. Occasionally, for variety, it slowly rotated a bit. Then it sat some more. If you are going to tell me that this "measures" an apparatus contraction somehow related to a quantity $\gamma = \sqrt{1 - v^2/c^2}$, I have to ask you what it is that moves at this speed "v" with respect to what. No independent provision in the experimental protocol involves the quantification of any such parameter. As I remarked in *Heretical Verities*,[2.11] the only thing characterized by v is an ether wind howling through the minds of the experimenters. MM was more an experiment in group psychology than physics.

I was recently asked to review a paper in which the author claimed that the Lorentz contraction of extended structures has been confirmed by hundreds of experiments. None was cited, but the possibility of such a claim demonstrates the existence of a pervasive myth among physicists. The metaphorical cemeteries are full of bushy-tailed young men who tripped into their laboratories eager to win renown as the first on their block to observe the famous Lorentz contraction and who ultimately emerged hollow-eyed husks, unsung and relegated to the teaching of freshmen. There have been possibly scores of ingenious attempts to observe such a phenomenon, all of which have failed and few of which have so much as earned mention by the textbook writers. Failure is bad news, which is no news to pass on. The most recent example of which I am aware that did achieve publication, is C. W. Sherwin, *Phys. Rev.* A **35**, 3650-3654 (1987). Like all before him, Sherwin failed to find a contraction effect. However, null experiments are tricky—there being always the possibility of effect cancellation by factors unanticipated by the experimentalist. (Recall the "mode locking" apparatus effect in the Sagnac example.) Consequently, neither Sherwin's nor the numerous other unsuccessful attempts to observe the Lorentz contraction can be taken as nulls of sufficient certitude to constitute proof against SRT (or against Lorentz's theory). But one can surely conclude that the unrelieved failure of attempts to achieve positive confirmation of this aspect of SRT leaves open the logical possibility of length invariance as a basis for alternative theory. That minor degree of concession must be asked of any physicists who would call themselves rational beings. The rest are either fanatics, or possessed of information I do not have, or both.

6.8 Clock rate as an energy state function

Our interest in "time" motivates us to examine in slightly more detail how it is measured. It will suffice here to confine attention to atomic clocks, specifically the cesium clocks used in the GPS.

The noteworthy thing not yet much emphasized about such clocks, or clocks in general, is the important property they possess of reversibility of changes in their running rates. That is, when work is done on a cesium clock to move it into a new state of motion or a new position in a gravity field, its natural running rate (rate of atomic oscillation v associated with a certain hyperfine-structure energy state transition $\Delta E_{trans} = hv$) is objectively slowed (provided motional state change dominates over gravity state change); but, if further work is done to restore the previous environmental and motional conditions, the clock is automatically objectively self-speeded in a dissipationless manner such that the former running rate is precisely restored. (I cannot prove this. It is a pure assumption, to be accorded axiomatic status.)

As for the cesium atoms themselves, we may model these as complex harmonic systems undergoing, in the case of motional changes, a total energy state change ΔE_{tot}, together with the reversible (oscillatory) internal energy state change associated with the atomic transition, ΔE_{trans}. I hypothesize that the γ-factor of time dilation objectively affects the aggregate of these terms, with the observable effect $hv \rightarrow hv/\gamma$ on the cesium clock's running rate. Thus the oscillation period $T = 1/v \rightarrow T\gamma$ increases, with the consequence of clock slowing. But this all occurs in what amounts to an atomic pure state; so, when the initial total energy state is restored (*e.g.*, by clock transport) the original oscillation period is restored without dissipative losses. This is plausible if the basic physical timekeeping entities, the atoms, remain uninterruptedly in quantum pure states—these being known to be dissipationless or energy-conserving. Consequently, clock rate may be considered a *state function*, as in thermodynamics. That is, it depends on state of motion and position in a gravity field, hence on total energy state. We may formalize this as a

> *Timekeeping Postulate 1.* Clock rate is a *state function* and is independent of motional history. It changes in a manner that continuously tracks the progression of a clock's total energy state changes. Restoration of state restores clock rate. All energy state changes may be referred to a given (arbitrary) initial state of inertial motion.

Thus, to predict how the timekeeping properties of any clock will change in passing from an initial to a final state, it suffices to know its running rate in the initial total energy state and its final total energy state. A specific candidate "law" quantitatively describing clock rate change dependence on (motional and gravitational) energy state change will be proposed in Chapter 8. There, the relevant descriptor will be found to be *action* rather than energy.

From this Postulate 1—which assumes no rate "hysteresis" of a clock repeatedly cycled through different environmental states—it is evident that *all clocks in a given state of motion and environment run at the same rate, regardless of their accelerational histories, provided they can trace (by transport) their ancestral origins to a common initial state.* This proviso can be dropped in the case of atomic clocks, if we specify our clock type in terms of a given atomic species and assume that species to behave reproducibly whenever reverting to an identical motional and gravity potential state. It hardly needs saying that all such physical postulates should be checked by physical experiments, wherever possible. In fact, as we shall see in the next chapter, there is room for doubt about the physical validity of Postulate I, so credence should be withheld until empirical evidence becomes available.

For many practical purposes clock rate information fully suffices, without concern for clock phases. Although the state function postulate settles (rightly or wrongly, in the manner of all postulates) the identity of clock rates in identical energy states, the question of clock phases—that is, of actual elapsed "time" (proper time or aging) told by a given uncompensated clock is more difficult, being path dependent. Here path history information is essential. (This accords with the inexact nature of the proper time differential $d\tau$.) As we shall see in the next chapter, the situation is entirely different for collective time t_0, told by rate compensated clocks. Here the differential dt_0 is exact and there is no dependence on path or history of motion. Once compensated clocks are synchronized, as by radio signals, they stay in synchronism regardless of relative motions (in general requiring, however, continual adjustments of rate compensation!).

As previously noted, it is sometimes suggested that acceleration is what controls the net aging of clocks and space travelers. Can less information than full accelerational history serve? Yes, indeed, less does clearly suffice—although the instrumentation required to obtain it would not be less. Consideration of Eq. (3.8a), *viz.*,

$$\tau_{traveler} = \int \gamma^{-1} dt = \int \sqrt{1 - v_{traveler}^2 / c^2} \, dt_{inertial} \, , \tag{6.14}$$

shows that in order to know all about a traveler's proper-time (uncompensated) clock reading, shown throughout every stage of his journey, it suffices to know the full history, not of his acceleration, but of his *gamma-factor*—which is to say, of v^2 relative to an inertial system. Such information—on the full history of v^2— appears to be both necessary and sufficient to describe the effect of motion on timekeeping at the phase level of accuracy. Here v is the relative speed of the traveler with respect to his starting inertial system. So, implicit in any complete accounting for the physics of natural timekeeping is an initial or "fundamental" system to which timekeeping reference is made (analogous to the rest system of the GPS Master Clock). It does not matter what starting inertial system is selected, but selected it must be. If, later, any other system emerges for some reason as more convenient for reference, a full history of v^2 for clock transport connecting these two candidate "fundamental" reference systems must be known … else no complete account of natural (proper time) timekeeping can be given.

That v^2, rather than velocity $\vec{v}$ or acceleration $\vec{a}$, is the crucial parameter for timekeeping can be seen from the CERN example.[6.3] The circling muons were continually changing their vectorial velocities $\vec{v}$ and accelerations $\vec{a}$, but maintained approximately constant v^2—and it is the latter that properly describes their observed uniform rate of aging. The role of accelerational history (if introduced) is simply to supply the information necessary to determine the v^2 history. If we know the history of v^2 changes, this is equivalent to knowing the history of kinetic energy changes. So, all this can be stated alternatively in terms of energy (or action) changes.

This circumstance, by an easy extrapolation, focuses attention on total energy changes as the physical gauge of clock rate changes—the best clue yet to a "causal" arrangement in nature. Indeed, we are here merely following in the footsteps of Ronald Hatch,[6.6] who stated explicitly that "It is the kinetic energy which determines the time dilation, not the acceleration." He similarly observed in connection with gravity effects on time, that " … it is not the acceleration which defines the time dilation, but the gravitational potential energy." Employing these hints, we shall attempt to develop the total energy (action) theme more fully in Chapter 8.

6.9 Reversible work

The reversibility of clock rate change postulated above is (if valid) a novel and non-trivial property of atomic timekeeping. The work involved, too, has a character of "reversibility" worthy of particular comment: In conventional thermodynamics "work" is path dependent and never truly reversible. In the closest approach to reversibility, it is pictured as an idealized, imaginary process typically associated with gradual, infinitesimal motions, such as piston travel, always close to a perfect equilibrium never fully realized in practice. In contrast, when applied to basic timekeeping at the atomic level, reversible work must be pictured as a *real finite process*; e.g., performed on a muon clock or a cesium atom clock in finitely (and possibly rather rapidly) altering its pure state of motional or potential energy. This must be conceived as a dissipationless process, such that on return to its initial pure state the clock returns to precisely its former running rate. In the case of motion, clock rate is at all times strictly controlled by γ, which starts from 1, increases beyond and decreases back to 1 on return to an initial state of motion.

Viewed in classical terms, reversible work seems to me at best paradoxical. Given that work depends on v^2, which is signless, the claim that doing positive work on a clock to move it with velocity $\bar{v}$ from state of motion A to state of motion B will cause it to run slow, whereas doing equal positive work on it to move it with velocity $-\bar{v}$ from B to A will cause it to regain its fast-running condition, seems paradoxical if not contradictory. Personally, I can accept this (if observation confirms that there is no alternative) only as another manifestation of the profound difference between the classical world and the world of quantum pure states—certainly as discordant with the teachings of macroscopic experience as the wave-particle dualism. Assis[7.4] has refined the Machian notion that distant masses of the universe are the physical "cause" of inertia—just as nearby masses are the cause of gravity. A parallel conception may apply to timekeeping; *viz.*, that distant masses (by instant action) "cause" the quantum pure-state clock to speed up or slow down with each change of motion state. Too spooky for you? For Nature, perhaps not spooky enough.

6.10 Atomic clocks: prospects for their improvement

There is little point in talking about time apart from atomic time, which has come to dominate the subject. (Pulsar time might at the current level of refinement give some competition.) It may be of interest to take a brief side excursion to examine the present art and its susceptibility to future improvement. One can quickly get an education in the basics of cesium beam clocks from the Internet, and I shall not repeat what can be learned there, apart from a brief summary. The most accurate present-day cesium clock is of the "fountain" type. As currently embodied, it requires the action of gravity to make it work. Applying radiation pressure, six orthogonal laser beams hit and cool a cloud of cesium atoms in vacuum, compressing it into a small ball. When hit with an upward-directed laser beam, the ball is projected upward through a microwave cavity filled with radiation tuned to the frequency v of the hyperfine transition that provides the fiducial timekeeping reference. It passes on through the cavity, reaches a maximum height above it, and falls back through the cavity under gravity, thus experiencing two successive resonant stimulations, separated by a time interval $\Delta\tau$ of about 1 second. This principle of "double-end tickling," due to Norman F. Ramsey[6.7] (developed by him for nuclear magnetic resonance studies of molecular beams), with a possible assist from Henry B. Silsbee,[6.8] lends itself to sharpening the resonance line in accordance with $\Delta v \cdot \Delta\tau \sim 1$; so, the longer the interval $\Delta\tau$, the better the line sharpening and the more accurately the clock can tell time. (The principle is applicable in many contexts, as in astronomy, where one can imagine a huge Newtonian parabolic telescope mirror with its reflecting surface entirely covered except for two small openings at opposite edges. The big mirror can then be discarded, leaving only the two small, far-separated mirror segments. These will lack intensity, but will offer much of the resolution of the big mirror. It is by application of a similar principle over thousands of miles of separation that the VLBI system attains its fabulous angular resolution.)

To improve future atomic clock accuracy, then, one approach calls for exploiting some variant of the fountain principle, with increased $\Delta\tau$, to achieve whatever line-sharpening (timing accuracy) we desire. Since gravity is needed to make the fountain work, but is turned off for a free-falling clock in orbit, it is evident that in an orbiting clock some gravity substitute must be employed for moving the ball of cesium atoms. Obviously, laser beams can serve this purpose, and can also be used to recompress the ball repeatedly, as need be to prevent excessive signal loss by diffusion. I can see no obstacle in principle to timekeeping accuracy improvements by factors of 10 or 100. Line sharpening can be improved (at a proportional cost in data rate) by progressively increasing $\Delta\tau$, aided in data processing by some convergence-speeding algorithm—possibly patterned on the "ε-algorithm" well-known to numerical analysts. Also it seems to me the prospects are good for miniaturizing the technology, certainly sufficiently for putting such clocks into orbit.

An alternative approach uses atoms other than cesium, having higher resonance frequencies. Success along this line has recently been reported using the mercury atom, with as much as 10-fold improvement in accuracy attained in even the earliest embodiment. What purpose will be served by clocks of ever-greater accuracy? Improved GPS positioning accuracy will serve evident and inevitable purposes, for good and evil. Beyond this, in the short term, possibly only research, including relativity research, will be furthered. But the advances of technology seldom for any length of time outstrip their ever-growing human uses.

6.11 Chapter summary

With reinstatement of the absoluteness of simultaneity, established through a corrected analysis of the train problem (based on description of light propagation by invariant Hertzian or neo-Hertzian electromagnetism, rather than by covariant Maxwell theory), comes a restoration of the present tense—of the concept of a physical "now." The separateness of all clock *rate* considerations from the question of *existence of a distant simultaneity* is to be noted. That existence fits perfectly with our postulated invariance of length, to which it is equivalent. Clocks can be running at various natural (proper time) rates in various systems and distant simultaneity will still be definable and "absolute," provided length is invariant.

GPS evidence is cited to support the objectively factual nature of clock slowing by a factor $\gamma(v^2)$ when the clock has ("reversible") work done on it to transport it into a state of relative motion described by v^2. In order for the GPS to function in the simplest way, it is necessary that clocks placed in orbit tell time at the same rate as earth-surface clocks. This is accomplished by pre-compensating the clocks to be placed in orbit—setting them to run *faster* than earth-surface clocks by the factor γ—so as to tell a "collective time" t_0 when placed in orbit. (Gravity potential difference corrections are needed, too, but I have simplified by deferring discussion of these to Chapter 8.) An uncompensated clock, when placed in orbit, is in free fall; hence it tells proper time in the orbiting local inertial system. Since all compensated clocks, each speeded-up by the appropriate γ-factor and placed in orbit, are observed empirically to run (count seconds) in step with the earth-surface clock, as intended, it follows that an uncompensated proper-time clock in orbit must run *slow* by the associated γ-factor (that is, slow compared both to the co-moving orbiting compensated clock and to the earth-surface proper-time clock with which the latter is in step) ... and this must be an objective fact. To repeat, the orbiting proper-time clock must run objectively slower than the earth-surface proper-time clock. There thus arises a basic timekeeping asymmetry, contradictory of the symmetry asserted by the form of relativity principle (*cf.* our "false theorem," above) promulgated by Poincaré and by Einstein. The γ-slowing that is only apparent according to that form of relativity principle is objectively real—hence asymmetrical between pairs of inertial systems—according to GPS evidence. The fact that some of the GPS clocks are supported in a gravity field while others are in free fall does not invalidate this evidence against Einstein's form of the relativity principle.

In this chapter we have postulated that similar clocks brought into a common state of motion at a common gravitational potential will run at the same rate, regardless of prior accelerational history. Clock rate hence becomes a *state function* (total energy or action state), and gravity effects as well as motional effects on timekeeping may be comprehended. Changes of total energy or action state thus directly correlate with changes of clock running rate. According to the view advanced here, the subtlety attributed to the *Herrgott* by Einstein is entirely an attribute of his own reasoning. Maxwell's equations, the "second postulate" of light-speed constancy, and even the relativity principle in its pristine form, all lie outside the *Herrgott's* plan ... as any sentient soul not blinded by subtlety can infer from GPS facts and others now long in the public domain. The deepest-lying and longest-persisting troubles of science have never stemmed from those refined subtleties that pre-eminently challenge the scholarly scientific mind ... always from gross misapprehensions unchallenged by it. These, like the systematic errors that plague the statistician, are hardest to identify and eradicate.

References for Chapter 6

[6.1]   A. Aspect, P. Grangier, G. Roger: "Experimental realization of Einstein-Podolsky-Rosen-Bohm *Gedanken* experiment; a new violation of Bell's inequalities" *Physical Review Letters* **49** #2, 91 (12 July 1982).

[6.2]   T. E. Phipps, Jr., "Meditations on Action-at-a-Distance," in *Instantaneous Action at a Distance in Modern Physics: "Pro" and "Contra,"* A. E. Chubykalo, V. Pope, and R. Smirnov-Rueda, eds. (Nova Science, Huntington, NY, 2001).

[6.3]   J. Bailey *et al.*, *Nature* **268**, 301 (1977).

[6.4]   R. P. Feynman, R. B. Leighton, and M. Sands, *The Feynman Lectures on Physics Definitive Edition* (Addison-Wesley, San Francisco, 2006), p. 16-3.

[6.5]   J. F. Taylor and J. A. Wheeler, *Spacetime Physics* (Freeman, San Francisco, 1966), p. 95.

[6.6]   R. R. Hatch, *Escape from Einstein* (Kneat, Wilmington, CA, 1992).

[6.7]   N. F. Ramsey, *Phys. Rev.* **78**, 695 (1950).

[6.8]   H. B. Silsbee, "The Radio-Frequency Spectra of H_2 and D_2 in Low Magnetic Fields," Thesis, Department of Physics, Harvard University, Nov., 1950.

The concept has not yet been found which describes simply the temporal relations of the universe.

—P. W. Bridgman, *The Logic of Modern Physics*

Chapter 7

Collective Time

7.1 Principles governing proper time

This may be as good a place as any to summarize my thesis in regard to Einstein's special relativity theory (SRT): As mathematics it is impeccable. As physics it constitutes a tightly-woven tapestry of fact (empirically validated) and myth:

Myth: Maxwell's equations tell the last word about electromagnetic formalism.

Myth: The Lorentz force law tells the last word about electromagnetic force.

Myth: Metric nature of spacetime.

Myth: Existence of spacetime symmetry (dictated by Maxwell's equations).

Myth: Universal constancy of c (dictated by Maxwell's equations).

Fact: Independence of light speed from source motion.

Myth: Independence of light speed from sink motion.

Fact: Existence of time dilation.

Myth: Existence of Lorentz contraction (dictated by spacetime symmetry).

Fact: Invariance of proper-time interval under general transformations. (Strictly, this applies to "alias," not to "alibi," transformations—but I omit discussion of this important distinction as excessively technical.)

Myth: Invariance of proper-space interval (dictated by spacetime symmetry).

Myth: Relativistic symmetry (reciprocity) of time dilation.

Myth: Velocity reciprocity as measured by naturally running (proper-time) clocks.

Myth: Validity of a strong relativity principle for environmentally-affected (naturally running, uncompensated) clocks.

Etc.

As a reminder, let me restate three of the basic premises of the alternative paradigm advocated in this book:

Premise (a): Hertzian or Neo-Hertzian (invariant) electromagnetism.

Premise (b): Length invariance.

Premise (c): Objectively asymmetrical alteration of the running rates of transported clocks under energy (action) state changes.

Einstein was not explicit about the principles he assumed to govern the natural running of (proper time) clocks in the absence of gravity effects. As nearly as I can discern these include what I called in the last chapter Timekeeping Postulate I, which I may rephrase as

Postulate I (*State function*). Clocks sharing the same state of motion all run at the same natural (proper-time) rate, regardless of acceleration history.

[Empirical evidence in support: I should judge that distant star spectroscopic evidence supports this at the atomic level, so it should apply at least to timekeeping by atomic clocks. However, such evidence is far from conclusive. There is no control over the "experiment."]

Corollary (*Reversibility*). A clock transported from one state of motion to another and then transported back to its initial state will resume its initial running rate.

[Empirical evidence in support: I would cite the Hafele-Keating experiment with round-the-world cesium clocks on the supposition that the returned clocks resumed their original running rates … except that, like the Eddington eclipse evidence that catapulted general relativity theory into media prominence, biased data selection[2.13] rendered the results highly suspect. More generally, it seems likely that enough experience has been gained with atomic clocks that if high-speed transport permanently and irreversibly altered their running rates in any systematic way, this would have become a recognized feature of the "art." Admittedly, this is inconclusive evidence of the "dog in the night" sort—*cf.* the Conan Doyle story "Silver Blaze." The stated Corollary represents a special case of Postulate I.]

Both of these results seem to be more or less taken for granted in ordinary presentations of SRT. It appears to me that Einstein may have gone farther in assuming a stronger form of Postulate I, to the effect that the natural (proper-time) running rates of clocks are independent of their state of motion (the same in all states) ... but that can have been only on Tuesdays, Thursdays, and Saturdays, when clock-slowing was a symmetrical "appearance" between pairs of inertial systems. On Mondays, Wednesdays, and Fridays, when clock-slowing was an asymmetrical reality, such a strong form conflicts with empirical fact (CERN, GPS, *etc.*, data), as reflected in our Premise (c). Since we shall maintain Premise (c), only the weak form of Postulate I, stated above, will be assumed here ... with reservations, since there is as yet little or no research to justify firm opinions in this area.

7.2 Collective time and relativity principles

SRT rests overtly on two postulates—relativity and light-speed constancy. Consider the first of these, sometimes formulated as:

> *Relativity Principle I* (Weak form): The laws of nature are the same in all inertial systems.

In applying this principle it is necessary to make a careful distinction between "laws of nature" (which the physicist understands to be general statements, typically having the character of descriptive differential equations) and specificities, which are particularizing data of the nature of initial conditions, boundary conditions, decay constant values, *etc.* A prototypical example of a "law of nature" is Newton's Second Law, which states that $\vec{F} = d\vec{p}/dt \rightarrow md^2\vec{r}/dt^2$. This differential equation makes no predictions about observables unless supplemented by initial-condition numerical values, known as "constants of the motion." Only through a combination of *general statements* and *specific input data* does the mathematical machinery of physics make numerical predictions of the results of observation or measurement. We need to be fully aware that the world as we experience it is a sequence of specificities, and that the relativity principle, in the form given above, does not address that sequence. According to the principle, as stated, it is only the "laws" that show invariance under inertial transformations. In other words *measurement results* can be expected to vary with the inertial system, not to show "invariance." Such variation is entirely in accord with the literal statement of the relativity principle in Form I.

Einstein went on to add:

> *Second Postulate*: The speed of light is measured in all inertial systems to have the same numerical value c (for time measurement by naturally-running clocks).

Thus one particular example of "specific data" about nature was elevated to a special position—light-speed c was elected to resemble a "law of nature" in obeying the relativity principle. That is, the constant c joined the exclusive club of invariants under inertial transformations. Now, this is very strange, for it completely blurs the line between two otherwise distinct categories of physical descriptors. If privileged constants can cross that line, why not all constants? And in fact that is essentially what happened. The two postulates endorsed the spacetime symmetry fostered by the partial differential operator symmetry of Maxwell's equations, and spacetime symmetry in turn produced complementary contortions of space and time measures—to keep c constant. As a result, not only muon decay constants, but separate measures of length and duration as well, became *in the co-moving or rest system* identical for all physical choices of that inertial rest system. So the relativity principle ceased to be valid merely for laws of nature and became valid for laws of nature plus specific numerical data relating to space and time mensuration.

Requiring the relativity principle to hold for that one constant c, in combination with the relativity principle itself, had the effect of making that principle seem to hold for more general aspects of physical experience. According to Einstein, the muon decays in 2.2 microseconds as measured (in terms of proper time) in every inertial system in which it is at rest. That the same muon is measured to decay much more slowly when moving in our lab is interpreted as evidence that "time" means different things in different systems. If, instead, we were to interpret time to mean the same thing in all systems, this would imply that different decay constant numerical values would have to be assigned in different systems. That would amount to no more than recognizing that motion affects decay rates (the "specific input data") rather than "time." Either way, Form I of the relativity principle is honored.

So Einstein was really presenting a more radical proposition. *In effect*, he was tacitly endorsing a different and stronger form of the relativity principle, which may be expressed as Form II:

> *Relativity Principle II* (Strong form): The numerical results of all measurements conducted according to identical operational procedures are the same in all inertial systems, when corresponding units are chosen, when the system (with measuring apparatus) is considered 'at rest,' and when clocks 'at rest' are specified to run at their natural (proper-time) rates.

Note that "laws of nature" are only implicit in this strong form of the relativity principle … explicitly, the *specific numerical results* of measurements conducted according to shared operational protocols are asserted to be *the same* in all physical inertial systems in which the observer places himself and his measuring apparatus 'at rest.' A superficial viewing of these two forms of the relativity principle might leave the impression that they are substantially the same. But this is not the case. Form II wipes out the distinction between laws of nature and the input data that enable such laws to make specific physical predictions. This second form is a more stringent assertion about the description of nature. Form II implies the validity of Form I. Thus Form I might be true and Form II not true. Or, they might both be true. In view of the considerations discussed in Chapter 6, Section 4 (where we observed that *"The operational procedures of clock compensation that work for clock transfers from earth to orbit do not work for identical transfers in the other direction"*) it appears that GPS evidence counter-indicates Form II. To make Form II true within the ambit of Einstein's paradigm, space (Lorentz) contractions must match natural time dilations in such a way as always to maintain the constancy, not only of c, but of other (separate length and time) measurements as well. That is, not only laws of nature but specific measurement data of all kinds must be unaffected by (*i.e.*, invariant under) inertial transformations. Whereas in Form I no such metric restrictions are implied. In discussing these matters below I shall try to avoid confusion as to which form of the relativity principle I am talking about. My point in emphasizing the distinction between the two is that Form I is weak enough to survive a re-parameterization of "time" in terms of collective time, whereas Form II does not in general survive.

The constancy of light speed, which is the key feature of Einstein's paradigm, can trace its ancestry only to Maxwell's equations, and must stand or fall with them. In my early chapters I touted Hertz's formulation [Premise (a), above] of the equations of electromagnetism as superior to Maxwell's. To the extent that I made a valid point there, the key feature fails—for light speed is not equal to the universal constant c in Hertz's theory. Next, to review my alternative approach, we recognized (in Chapter 6, Section 1) that Hertz's equations are compatible with the absoluteness of distant simultaneity. Absolute simultaneity in turn implies the invariance of length (as was argued in Chapter 6, Section 7), which is Premise (b), above. Finally, our Premise (c) challenges the possibility of a universal c-value measured by naturally-running clocks. It denies the "relativistic symmetry" of rates of such clocks—asserting instead a genuine physical rate asymmetry of high-speed transported proper-time clocks. In effect, with such clocks, to predict their rates one needs to keep track of their energy state changes.

Let us turn now to the consideration of collective time:

> *Definition of Collective Time (CT)*: CT is the form of time measured in principle by a space-filling collection of arbitrarily-moving clocks, each of which has been transported from a state of co-motion with an inertial Master Clock and pre-compensated in running rate in such a way as to anticipate and remove any rate-influencing effects of altered gravitational or motional environment. The clocks of the resulting collective all run at the same rate and are kept permanently in step with the Master Clock by further corrections tailored to the individual needs of each transported clock.

We find CT in general incompatible with Form II of the relativity principle, as noted, but compatible with Form I. Although physical process rates vary with state of motion—so that the rate of flow of collective time is in general dependent on choice of the Master Clock's ("fiducial") inertial rest system—the "laws of nature" do not depend on that flow rate. An alternative expression of Relativity Principle I is therefore: *All inertial systems are equally suited to serve as fiducial systems for expressing the laws of nature in terms of CT.*

> *Consistency Theorem:* Relativity Principle I, in the alternative form just stated, is valid under CT.

Proof: Consider inertial system A as the chosen fiducial system (rest system of a Master Clock) and system X_i, a typical clock-transport destination system, which need not be inertial. Rate-slowing effects of clock transport from A to X_i, dependent on the scalar parameter v_i^2 (where v_i is the relative speed between A and X_i), are compensated in the way already discussed; viz., by taking $N'_{0X_i} = N_0/\gamma_{AX_i}$ for the transported clock, where $\gamma_{AX_i} = 1/\sqrt{1 - v_i^2/c^2}$. Gravity effects are independently compensated on the basis of potential energy state change, determined by the geometry of the gravity field (as will be discussed in Chapter 8). This is the method referred to in the CT definition. After application of this method of compensation and after transport from system A to each of any number of differently-moving systems X_i, each of a corresponding number of individually transported clocks tells collective time $t_A = t_0$, in rate synchrony with each other and with the Master Clock at rest in system A.

Suppose instead that an entirely independent establishment of CT is undertaken, the Master Clock being placed at rest in a different inertial system B. Similar operational procedures lead to apparently inconsistent results. For in this case, in which B is the fiducial system (B playing the role previously assigned to A), we cannot expect the resulting synchronizations, carried out by operational procedures analogous to those employed when A was the fiducial system, to produce a time flow rate characterized by $t_A = t_0$, because the compensation factor γ_{BX_i} will in general differ numerically from the previous factor γ_{AX_i}. All clocks transferred from B will indeed end up rate synchronized, but their *absolute rate of running* will not be measured by $t_A = t_0$ but by something proportional to that, say, $t_B = \alpha_B t_0$. That is, for different choices B of the fiducial inertial system all compensated clocks will in every case after transfer run at uniform rates in mutual synchronism, but those rates will vary with the choice of B. The resulting rates (for B as fiducial system) will bear to the original rate (for A as fiducial system) a proportionality factor $\alpha_B = t_B/t_A = t_B/t_0$. The nominal time flow rates in A and B differ numerically, consequently the measure of process evolution rate differs, but the difference is a simple one of proportionality, so that any chosen initial time flow rate associated with t_0 in A can be recovered *numerically* in arbitrary inertial system B by applying a constant multiplier, $t_0 = \alpha_B^{-1} t_B$.

Since the α_B-values vary with choice of B, it is apparent that we lose a strong relativity principle, understood in Einstein's Form II sense of *equivalent measurement physics* in all inertial systems—so that different inertial systems do not serve as "equivalent" fiducial systems. But we verify next that the weaker formulation of the "relativity principle," Form I, allows it to hold. To recall, Form I merely asserts the invariance of "laws of nature" under inertial transformations. Understood in this less restrictive way, a relativity principle is not lost for several reasons, three of which may be enumerated as follows:

(1) We recognize that in nature there is no such thing as a numerically measurable or conceptually meaningful *absolute* rate of "time flow." *Time itself*, being a metaphysical entity, is not directly measurable (but has to be inferred, with the help of convention, from observation of physical repetitive processes). Hence descriptive schemes that treat time flow rate as indeterminate within a numerical multiplier are merely reflecting the way nature is. In other words, descriptions that are the same, *modulo* a multiplicative factor on the time flow rate, are "equivalent" in a formal or abstract mathematical sense—so the clock rate discrepancy between transporting rate-compensated synchronized clocks from A to X_i and transporting them from B to X_i reflects no discrepancy of the "laws of nature" in the two systems A and B.

(2) Another way of interpreting the numerical indeterminacy of the "laws of nature" is to observe that such laws do not depend formally on choice of measurement *units*. (*Cf.* the Principle of Similitude mentioned below.) Allowing for adjustments of time units, the laws of nature are the same regardless of the paths chosen for synchronization of transported clocks. Our clock rate proportionality factor α_B can thus be interpreted as a time units factor.

(3) Finally, if it is important to do so, we can formally enforce $\alpha_B = 1$ for all B through a *two-stage compensation* procedure: The first stage produces in general $\alpha_B \neq 1$, as above,

and the second stage is just a further numerical adjustment, applicable uniformly to *all* clocks, including the Master, to establish the use of a time unit consistent with some pre-ordained *shared* standard. Thus there exists in principle a (double-jointed) compensation scheme that not only synchronizes all clocks, but causes them to register the same t_0 (same nominal time flow rate), independently of choice of physical fiducial system. For such a specialized compensation scheme the strong Form II of the relativity principle (with CT timekeeping instead of natural timekeeping) is presumably obeyed. From this observation we infer that the form and even the possibility of a relativity principle is more sensitive to the operational details of timekeeping than might have been thought.

This last does not say that muons, after two-stage compensation, will decay at the same rate in all systems ... rather, that the "second" will mean the same thing in all systems. With reference to different fiducial systems muons will necessarily be described by numerically different decay constant values. Decay "constants" are not laws of nature, nor are they in this context constants. Indeed, it seems inexpedient to conceive of them as constants ... much better (closer to the physics) is to accept that environment and motion are affecting decay rates rather than affecting "time." For example, an observer co-moving with the CERN high-speed muons and using a compensated ($\gamma = 29$-fold speeded-up to keep it in step with the lab-stationary clock) CT clock measures the transported muon half-life, which has been objectively lengthened $\gamma = 29$-fold, as 2.2×29 microseconds of CT (numerically equal, of course, to what the CT Master Clock at rest in the laboratory measures for the same moving muon); whereas, using an uncompensated clock, he measures it as 2.2 microseconds of proper time, because his proper-time clock runs 29 times slower than his CT clock and thus registers 29 times less elapsed time. So the moving observer has to use a different decay constant, depending on whether he consults his CT clock or his proper-time clock.

The physics of this can perhaps best be comprehended by hypothesizing a "sympathy" or similitude between muons and the cesium atoms of an atomic clock, such that the muon half-life corresponds to the same number n of cesium oscillations, regardless of the environment they may share. When they are placed together in a state of high-speed motion or strengthened gravity field, the cesium oscillations are objectively slowed. As a result, the muon half-life is objectively extended, as a matter of physics, to match the same required number n of the slowed cesium oscillations. This fact of slowing of the oscillations is hidden for measurement purposes, if the cesium clock is *uncompensated*, by the fact that its proper-time "second" is defined as a given number N_0 of oscillations that bears a fixed ratio to n, the same ratio in all environments; so the same number of microseconds (2.2) is "measured" in all environments by a (proper-time) clock always co-moving with the muons. But, if the high-speed co-moving clock is *compensated*, its "second" is defined as a reduced number $N_0' < N_0$ of oscillations; so more of these seconds elapse during the n cesium oscillations needed to complete a half-life. Consequently the "measured" (in terms of seconds) half-life under CERN conditions of high-speed motion is increased ($\gamma = 29$-fold), for measurement by means of CT. The result is that in the CERN experiment both the lab-stationary clock and the muon co-moving CT clock measure a half-life of 29×2.2 microseconds. This agreement of clocks in different states of motion allows the half-life extension *caused by* the work done in producing relative motion to be viewed as an objectively real phenomenon, to the extent that CT is accorded physical significance—in agreement with the premises of the present analysis. In other words, CT may be more than "just a convention;" but that is a long extrapolation from what can be proven on present data.

Einstein's Form II (strong form) of the relativity principle, asserting identical outcomes of measurements made with naturally-running uncompensated clocks, is empirically unverified, and is supported by nothing thus far observed (the *laboratory physics* for fast and slow muon decays being distinct and in no sense "equivalent"). However, the original Form I of relativity principle, when referred to time as measured by compensated collective-time (CT) clocks, is plainly confirmed and obeyed ... since the running rates of the high-speed and low-speed clocks are in a constant ratio α_B (about 29 in the CERN experiment[6.3]), and we can compensate all clocks in the universe to match either rate, simply redefining the "second" or N_0-value, as needed, without affecting the "laws of nature." We ordinarily ordain our lab second as the right one ... but there is only convenience or politics, no physics, in such an edict. The slow-decaying high-speed muon's stretched-out "second" might be chosen instead. We should have to recalibrate our clocks, redefine our "years," and get used to our centenarians living only 100/29 years, and the oldest dogs only $100/(7 \times 29)$, as well as our birthday intervals and trips around the sun taking only $1/29^{th}$ of a year. That might prove at first unpopular, but in America the annually-confirmed willingness of *vox populi* to support "daylight saving time" suggests that politics could accomplish it (except in western Indiana) with no disturbance of the laws of nature and no observable penalty to the physical life span of man or dog (in terms of trips around the Sun).

A different way to recognize the validity of Relativity Principle I under CT parameterization is to consider in terms of same-t_0 parameterization a postulated instantaneousness of distant force actions. Such instantaneousness is obviously *independent of choice of the Master Clock* and its state of motion, since an "infinite speed" of action can on a CT-speed scale be considered "equally distant" from any finite relative speed; so instantaneousness "looks the same" as viewed by observers in all finite states of motion. In other words, with reference to CT-speeds, anything possessed of "infinite" speed has "the same" speed relative to any finite speed. A force judged instantaneous with respect to one inertial system will therefore be judged instantaneous with respect to all. This is another way of expressing the viability and *universality* of a CT concept of "now."

The following fantasy may prove helpful in understanding CT: Suppose our solar system contains n planets, the inhabitants of each of which decide to create their own private GPS. The resulting n superposed GPS systems, when universalized, may then provide the basis for n different realizations or representations of collective time. Since the planets are polyglot, there is zero probability that any two of them use exactly the same definition of the unit of time. Consequently the running rates of these n CT's are all different. If a slave of one GPS were mistaken for the master of another, of *vice versa*, total confusion and dissention would result—perhaps a *casus belli*, in the pattern of Swift's wars of Big-enders and Little-enders. But, viewed disjointly, these n systems each furnish a (self-)consistent CT. Each planet's inhabitants can use their own system's CT to describe the "laws of nature," and those laws will be identical across all planets and systems … so a form of the relativity principle is honored. Note that there is no shortcut, no cheap solution: Each planet must ante up the cash, buy and expend the rocket fuel, perform the physical operations, to create a flock of orbiting slave clocks. Mental operations will not do it. One cannot promote a slave to master by just switching names or labels. This is doubly apparent from the fact that slaves can in general be in non-inertial states of motion, whereas masters are always inertial.

In summary: Natural physical processes proceed in different inertial systems at objectively different rates; thus the strong Form II of the relativity principle (based on "time" as told by naturally-running, environmentally-affected clocks) should in experimental practice be observed to fail. In fact, if (as here anticipated) the Lorentz contraction does not occur, it will necessarily fail. But if the relativity idea is expressed as in the weaker Form I, at its level of maximum abstraction in reference to "laws of nature," then our proposed GPS-analogous clock compensation method introduces a collective time t_0 entirely compatible with that form of the relativity idea. This completes our proof that it is possible to establish throughout the universe a collective time consistent with a form of the relativity principle.

7.3 Related observations

Remark on relativity principles. We need to emphasize a broad conceptual recognition concerning relativity principles in general. This is that they are not self-evident and not independent of our descriptive choices. Their validity is contingent on the details of time parameterization; for those details control the *simplicity of form* of the "laws of nature." There are ways to parameterize time that make that form complicated and ways that make it simple. Simple forms are more likely to have simple transformation properties ... and the form *invariance* asserted by a relativity principle is the simplest possible. Choose the wrong parameterization of time (as Einstein did) and you do not invalidate "laws of nature" but you may complicate their expression sufficiently to invalidate one or another form of the relativity principle. And you may lose the simplest transformation property of invariance and have to invent a more complicated version of it, such as covariance.

Remark on clock compensations. We have mentioned that the GPS engineers make clock running-rate corrections for unwanted effects of relative motion and position in the earth's gravity field. Other effects requiring compensation that have not been discussed here include a GPS "Sagnac effect" due to earth rotation, which was eliminated by choosing a *notional* (as distinct from a real) Master Clock at rest in a non-rotating fiducial "inertial" system co-moving with earth's center. Such a system is "sufficiently" inertial for their day-to-day or *diurnal* timekeeping purposes, but would not, I daresay, be sufficiently inertial for studies such as stellar aberration, an *annual* phenomenon. Kelly[2.13] differs from my opinion on this, in that he recognizes *no* annually periodic effect on GPS timekeeping. In this connection it might be noted that when the earth is at perihelion (closest to the sun) *all* earth and GPS clocks should in principle run slower, because of deeper penetration into the sun's gravity field and also because of greater orbital speed. Similarly all clocks should run faster at aphelion. But I have not verified that current timekeeping accuracy warrants such a prediction, the earth's orbital eccentricity being quite small. Pulsar timing might provide a check.

Remark on units. Perhaps in view of the key nature of the topic a word more should be said about *units*. The equations that express the laws of nature, as indicated above, do not in themselves suffice for physics. They contain symbols that have to be related to numbers ... and numbers can be assigned to symbols only after units of measurement have been verifiably determined. This is a matter of definition. I am indebted to Victor M. Waage for pointing out that Newton (*Principia* II, Proposition 32) gave the first discussion of the Principle of Similitude, according to which the form of a physical formula or equation is not altered when one makes alternative choices of values for the fundamental units of length, time lapse, and inertial mass. The choice of units is thus arbitrary and independent of "laws of nature." It is neither prescribed nor conscribed by a relativity principle governing such laws, but in each inertial system engenders a separate problem typified by that of "axis calibration." Einstein provided the world with a set of pristine inertial frames, in each of which the laws of nature were postulated to be the same. But he was not careful about how the axes of those frames were to be calibrated. It is necessary that a shared set of conventions be adopted in order for a common set of equations to be able to embody the laws of nature. How are hypothetical inhabitants of the hypothetical inertial frames to know what is meant by "meter" and "second"? We might wish to broaden this question to: what meanings should be attached to those terms in order to produce the simplest physical description?

The easiest answer (other methods amounting to the same thing) is that a physical transfer of *metric standards* may be considered to take place. These play a crucial role, little emphasized in the SRT literature. We claim here that physical length is invariant. If so, there is no problem about transferring standard meter sticks to define the meter. In principle two "particles" one meter apart and transferred *via* identical protocols of *simultaneous* acceleration will serve as well as any material standard. The "distant simultaneity" required to lend substance to this concept has been rehabilitated in Chapter 6. A corresponding material metric standard, in order to remain *stress-free* during any change of its state of motion, would require application of equal forces front and back, corresponding to the above-mentioned identical histories of simultaneous acceleration of the front and back particles. The reason is that Newton's physics applies at first order, and it asserts that any difference of force applications front and back will stress the material. Since there is no way high speeds can be reached without passing through low speeds (and since there are always co-moving inertial systems in which speeds are "low" and Newton's first-order physics applies), the requirement to avoid stress at all speeds forbids the onset of differences of front and back force applications at any speed.

SRT sees the matter quite differently. According to the claim of that theory—implausible according to Newton's first-order physics—the worldlines of the two ends of a meter stick that is set into longitudinal motion must be *differently curved* in order for that stick to remain a stress-free metric standard. This difference of curvature must arise from the very onset of motion, at zero speed in the heart of the Newtonian physical regime. In other words different forces must be applied to front and back—which would stress the stick at first order—so as to keep it stress-free at higher order. Go figure what kind of minds think that way! (They are the same minds that rule the editorial offices affiliated with the American Institute of Physics, responsible for a blanket interdiction on questioning of any aspect of SRT in AIP publications.)

Meter sticks may, as I assert, be a simple matter ... but clocks are another matter entirely. They are inherently more complicated and are found empirically to be more sensitive to environment. The meaning of a time "standard," particularly one that is to be transported among different physical environments, is therefore dubious. We have asserted in Chapter 6, have postulated, and have cited substantial empirical evidence for believing that any change of total *energy state* of a clock—whether gravitational or motional—affects its running rate ... somewhat as temperature change affects the length of the (former) standard meter in Paris. If we desire a common, invariant meaning of the meter and the second to be shared throughout our universe of discourse, we have to be ready to make allowance for any and all such un-wanted "apparatus effects" (*i.e.*, to compensate them by adjust-ments that are essentially definitional in character), including possibly effects not yet discovered. A standardization of meaning of the symbols representing length and time in our equations is unquestionably desirable in order to give *simplest expression* to the laws of nature. Indeed, the simplest expression may be essential to the *possibility* of formulating a valid relativity principle ... for we cannot expect arbitrarily complicated mathematical represen-tations to express simple symmetry properties of nature.

The desirability and feasibility of reducing laws of nature to simplest form has been part of the standard canon of physical de-scription ever since Newton's landmark recognition that the equations of mechanics simplify in inertial systems. Einstein's counter-recognition, that by inordinately complicating the mathematical expression of the laws of nature it is possible to eliminate the special status of inertial systems—so that a real as-pect of the world can be banished by coordinate juggling in men-tal space without need to comprehend the *physics* of inertial forces, or even to acknowledge their existence—is distinctly not a theme to be exploited in the present book. We may safely leave such self-flagellation for the jaded scholarly appetites of the fu-ture. Let's not rush into that sort of future ... rather, let's rejoice that we aren't there yet. If we get the simplicities right first, they may prove to hold unsuspected and rewarding elements of sur-prise, which will spare us untold mathematical masochism.

7.4 Philosophical context

As has been indicated at several places throughout the previous text, on a philosophical level we are here undertaking to restore what might be termed Platonic notions of ideal time and length, as distinguished from what clocks and yardsticks are doing in given environments. This "turns the clock back" from some of Einstein's (or Bridgman's[7.1]) deepest apperceptions. It is all very well to focus upon instruments and to stake everything upon the "measurements" they make, but nobody in his right mind would insist on ignoring known instrumental shortcomings—equivalent to playing on "a cloth untrue with a twisted cue and elliptical billiard balls" ... although four-index tensor symbols could undoubtedly perform the latter trick well enough for government work. It is my assumption that the known environmental susceptibilities of clocks may well (and even conservatively) be conceived in this context as instrumental shortcomings in need of compensation.

When a metric standard *measures* a change in the environment, there is ambiguity as to whether the standard changed or the environment changed. This ambiguity deepens whenever the possibility arises that the standard is part of the environment—that is, equally affected by whatever causes the change. Since the standard is itself created by a definition, the implication is that "compensation" for such effects of underlying physical variability, when they are unwanted, will require changes of definition. It is such definitional instability that I believe to be at work in the metrication of time—as in the GPS timekeeping definitional compensation $N_0 \to N_0'$. Still, I recognize that mankind, particularly *homo scientificus*, clings more lovingly to its definitions than to its gods. So I do not look for ready acceptance of the idea of sacrificing definitional stability to achieve conceptual or parametric stability.

The pendulum of fashion has swung very hard in our day toward castigating as metaphysics (in the pejorative sense) any view of time as distinct from what naturally-running clocks measure. But the people who feel that way about "time" have no hesitation in making a distinction between what their thermometers read and what the "temperature" is. They would not label such a distinction metaphysics, but would recognize it as the essence of sound physics—the kind that persistently seeks simplest descriptions. If (as is the case) it results in the simplest formulation of temperature-related "laws of nature," such as those of thermodynamics, to make compensations of thermometer readings before treating them as *temperature*, it is possible that a simplest formulation of time-related sciences such as mechanics and electromagnetism would result from the making of compensations of clock readings before treating them as *time*. That is the thought under development in this chapter.

Through clock-rate compensation, the "second" can acquire a universal meaning (*modulo* an arbitrary units multiplier), so that it becomes useful for ordering events everywhere and from all viewpoints. "Time" is then restored to its ancient intuitive meaning, conforming to personal experience, with well-defined past, present, and future, about which all variously-moving observers agree. One form of the relativity principle is honored, as we have just seen, and the equations of physics, crafted to express the "laws of nature," become valid in all inertial systems (which is to say invariant, as in Newton's physics—not covariant). The present chapter is devoted to exploring this idea of what we have called *collective time*, achieved through clock compensation, and possessing all the practical attributes of a universal time. Possibly ... not surely, because the theoretical pudding stands as always in need of empirical proof.

I confess that I have sometimes referred to my own version of operationalism as "instrumentalism" ... but I hope never to be confused with a purist of any kind. If I am going to place my trust in instruments, I have every right to be tough-minded in my demand for independent evidence of their trustworthiness. Where it is known from independent evidence that motion and gravity affect clock running rates, trust would seem to be a highly dubious, even over-presumptuous, commodity. True, it is not impossible that "time" ought to mean what mortal twins and muons seem to say it means—that it runs slow at high speed or in strong gravity. But this chapter is dedicated to showing that Newton's perception in this regard may well have been sounder than Einstein's, in the sense that it may in the long run prove more *fruitful* for simplifying the formalism, equations, and calculations of physics to think of time essentially as Newton did—as something by convention uniformly unfolding, undisturbed by environmental conniptions. Fruitfulness is not an aspect of physical description that physicists (even in their degraded guise as mathematicians *manqué*) can afford utterly to ignore. For even the purest of mathematicians (setting aside those of the hardened Hardy[7.2] variety) within their own discipline recognize fruitfulness as a criterion of "value."

7.5 Particle mechanics, again

I acknowledge that my proposed clock-rate-compensation approach to simplifying the description of nature appears so underhanded that many physicists will shrink from it instinctively as some form of moral turpitude. Can it be made more palatable? It seems indeed "against nature" to defeat the natural tendency of a clock to run slow when placed in a state of more rapid motion — but equally it is against nature to defeat the natural swelling tendency of a platinum meter bar in a hot room by keeping it artificially refrigerated. A society that consistently allowed moral inhibitions to stop it from defeating the tendencies of nature would freeze in the dark.

Further, consider this argument: The lion's share of the physical descriptive problem is addressed by one form or another of the science of mechanics, which describes the relative motions of objects and the forces or energy changes responsible for those motions. If we can certify compatibility of point particle mechanics with the suggested clock-setting scheme, a giant step will have been taken toward justifying confidence in it. But this is trivially easy. We have already done the work in Section 4.8. Let's reprise it in this new context.

Symbolize by t_0 the collective time told by clocks compensated in the manner we have been discussing. Its formal properties are those of any other "frame time," the frame being that in which the Master Clock is at rest. Since simplicity is our game, and mechanics is its name, we may confine our discussion to physical inertial frames. This means that t_0 is related to the (uncompensated) proper time τ of a particle moving arbitrarily in the fiducial inertial rest frame of the Master Clock by

$$d\tau = \frac{dt_0}{\gamma_0}, \tag{7.1a}$$

in agreement with Eq. (3.3b) and equivalents given elsewhere. From this follows

$$\frac{d}{d\tau} = \frac{dt_0}{d\tau}\frac{d}{dt_0} = \gamma_0 \frac{d}{dt_0}, \qquad (7.1b)$$

in agreement with (3.5). Here $\gamma_0 = 1/\sqrt{1-\left(v_0\left(t_0\right)/c\right)^2}$, where $\vec{v}_0 = \vec{v}_0\left(t_0\right) = d\vec{r}/dt_0$ is particle velocity measured in the inertial frame of the Master Clock, and $v_0 = \sqrt{\vec{v}_0 \cdot \vec{v}_0}$. The *invariant* (proper time) equation of particle motion, Eq. (4.26), generalized from Newton,

$$\vec{F}_{inv} = \frac{d}{d\tau}\vec{p}_{inv} = \frac{d}{d\tau}\left(m_0 \frac{d}{d\tau}\vec{r}\right), \qquad (7.2a)$$

depends only on parameters descriptive of the particle's location and intrinsic properties, so it holds in all circumstances (and recommends itself entirely independently of, and distinctly from, formal "covariance" considerations). In order to see how the particle moves as a function of collective time t_0, we have only to employ (7.1b) in (7.2a). Thus

$$\vec{F}_{inv} = \gamma_0 \frac{d}{dt_0}\left(m_0\gamma_0 \frac{d\vec{r}}{dt_0}\right) = \gamma_0 \frac{d}{dt_0}\left(m_0\gamma_0\vec{v}_0\right). \qquad (7.2b)$$

As with all mechanics expressed in its simplest form, going back to Newton, the particle position vector $\vec{r}$ has to be referred to an inertial system. The "physics" behind the simplification of mechanics in inertial systems, though controversial, is probably a reflection of the physicist's decision arbitrarily to separate off a "local" system for descriptive convenience from the rest of the universe. Hoyle and Narlikar[7.3] and also Assis in his "relational mechanics"[7.4] suggest that the right way to eliminate this special status of inertial systems, if the game is judged worth the candle, is to take account of the whole universe. In the present book I make no attempt to follow these and other pioneering authors such as the Graneaus[7.5] in their quest for a deeper understanding of nature's mechanism behind inertia. Instead, I content myself with the single-minded pursuit of a *simplest description*. In harmony with that objective, for physically defining inertial motion—although it is often convenient for practical purposes to think of the reference system as the "earth's surface"—it would be more accurate to use the barycenter of a non-rotating earth (as does the GPS system), or the barycenter of the solar system (for the description of stellar aberration), or even the barycenter of the galaxy (for some conceivable very long-term descriptive purposes, as mentioned in Chapter 2, Section 8). In the present instance this consideration applies to the "inertial" rest system of the Master Clock. All such reference systems might be superseded by a "centroid of the universe" as the ultimate inertial referent, if such proves operationally definable.

From (4.27) we have

$$\vec{F}_{inv} = \gamma_0 \vec{F}_0 , \qquad (7.3)$$

where we identify our previous "laboratory" with the inertial reference system under present consideration (the rest system of the Master Clock), wherein what might be referred to as "frame-time force" is now denoted $\vec{F}_0$ (force measured in the Master Clock's rest frame). From (7.2b) and (7.3) we obtain

$$\vec{F}_0 = \frac{d}{dt_0}(m_0 \gamma_0 \vec{v}_0), \qquad (7.4)$$

which is just Newton's law of motion, Eq. (4.25), modified by an effect conveniently interpreted as either "mass increase" ($m = m_0 \gamma_0$) or "velocity increase" ($\vec{V}_0 = \gamma_0 \vec{v}_0$). How do we know that such a simple force law holds, with instant action-at-a-distance, third-law force balance, etc., as in classical theory? We don't "know" it. It's an hypothesis, to be tested empirically with scepticism as always. Man proposes, nature disposes. Eq. (7.4) defines a collective-time mechanics that is obviously a covering theory of classical Newtonian point particle mechanics, but is known to be valid also at higher particle speeds where Newton fails.

All we have done here is to start with a formally invariant version of the mechanical law of motion, (7.2a), and strip it back to its Newtonian "frame time" origins, with automatic accommodation of "relativistic" mass increase. None of the transformation-theoretical machinery of SRT has been used. The differential dt_0 is both exact and invariant under general coordinate transformations (because our clock compensation scheme works for clock-particles in arbitrary states of motion). Exactness of dt_0 implies that, although we have here considered only the one-body problem, there is no impediment to generalizing the formalism in an obvious way to accommodate *many bodies* in simultaneous interaction (with Newtonian third-law action-reaction equality)— simultaneity being now well-defined in terms of collective time. By contrast, as long as one's intellectual horizon is limited to a proper-time formulation of mechanics [Eq. (7.2a)], such many-body generalization is next to impossible because of the inexactness (non-integrability, path-dependence) of $d\tau$, not to mention the loss of action-reaction balance. Such limitation and loss are inherent in the Einstein view of time as necessarily that which is told by naturally-running (uncompensated) clocks. When you have the sky full of differently-moving clocks, you have it full of differently-running proper "times." OK, if you must play it that way ... but why close your mind to alternatives?

Finally, we remark that Hoyle and Narlikar[7.3] go about eradicating field theory (an eminently worthy objective) by replacing the "field" with direct particle-particle interactions. Obviously, such interactions should obey Newtonian action-reaction balance (as Assis has it[7.4]); but instead, blinded by the science of spacetime, these otherwise innovative authors propose (for the sake of "relativistic invariance") to have their interactions conveyed by a "biscalar propagator" that depends on the whole worldlines of all particles comprising the rest of the universe. That is, the entire past and future of these distant particles are important for what they are doing to particles in the here and now. I suggest, with the resurgence of an omnipresent "now," that we allow such propagators to collapse onto it—with an audible sigh of relief.

7.6 Field theory revisited

We have just looked at mechanics and found it readily adaptable to a collective time formulation. The other great, if ruinously crumbling, pillar on which physics rests these days is field theory. What form does electromagnetic field theory take if all clocks are compensated to read collective time t_0? Our considerations favoring an invariant formulation (invariant under inertial, which is to say Galilean, transformations) remain valid, as in our discussion of Hertzian electromagnetism. In fact, the Hertzian formalism of Chapter 2, there valid only to first order, here becomes valid to higher orders, requiring only formal replacement of frame time t, wherever it appears in the field equations, with the collective time parameter t_0. The relationship of collective time, which is a *particular kind of frame time having invariance properties*, to a form of the relativity principle has been discussed in Section 7.2. In short, when an invariant (*note*: invariant, not covariant!) variety of frame time is taken as the fundamental time descriptor, there is no room or need for higher-order "corrections." (The invariance properties of collective time will be discussed in Chapter 8, Section 3.)

The Hertzian field equations (2.4) and the Hertzian wave equation (2.18), with t_0 for t,

$$\nabla^2 \vec{E} - \frac{1}{c^2}\frac{d^2}{dt_0^2}\vec{E} = 0,\tag{7.5}$$

hold, the latter with phase velocity solution (2.24), viz.,

$$u = \pm c + \frac{\vec{k}}{k}\cdot\vec{v}_d,\tag{7.6}$$

as before—the time parameter employed in calculating u, c, and $\vec{v}_d$ being understood to be collective time t_0. An equation similar to (7.5) holds, of course, for the $\vec{B}$-field.

We concluded earlier that the Hertzian account of electromagnetism is physically deficient, in that it fails to account for stellar aberration. Does this refute our present approach? Let us look more closely at this apparent difficulty. Collective timekeeping depends on a clock-compensating procedure that must be unravelled in order to find out what is predicted observationally. The unravelling will necessitate referring to proper time. If we confine attention to motional effects, a Rosetta stone for translating between compensated and uncompensated (proper-time) timekeeping is Eq. (3.3a),

$$\frac{d\tau}{dt} = \frac{d(proper\ time)}{d(frame\ time)} = \gamma^{-1} = \sqrt{1 - v^2/c^2}.\tag{7.7}$$

Here "v" is what we identified as the velocity, relative to an inertial frame as fiducial referent, of the point "detector" particle whose proper time is involved; that is, $v = v_d$. In the case of stellar aberration, v_d is earth's orbital velocity relative, say, to the Sun (better, the barycenter of the solar system) as fiducial referent. In the present problem collective time t_0 can be considered to play the role of frame time of this fiducial inertial frame, and the time t_{obs} of an astronomical observatory fixed on the earth's surface can be considered to play the role of proper time τ of a point "detector" or telescope in motion at relative speed v_d in the fiducial frame. Alternatively, t_0 can be the frame time of an inertial system instantaneously co-moving with the telescope at one moment, and observatory time t_{obs} can be the proper time of the telescope at a later moment when the relative velocity of the two is

$\bar{v}_d$. (In the latter case it is *change* of aberration angle that is described.) In either case since we have in (7.7) $\tau \to t_{obs}$ and $t \to t_0$

$$\frac{d\tau}{dt} \to \frac{dt_{obs}}{dt_0} = \gamma^{-1} = \sqrt{1 - v_d^2/c^2} \ . \tag{7.8}$$

Let us consider a particular instance of stellar aberration—namely, that of starlight coming straight down from the vertical (normal to the ecliptic plane). For this case $\bar{k} \cdot \bar{v}_d = 0$, as in Chapter 4, Section 3. As we said, the phase velocity equation (7.6) refers all its velocities to collective time t_0. Consequently in that equation the speed of light $u = \pm c$ is modified to

$$u = \pm \frac{dx_{phot}}{dt_0} = \pm \frac{dt_{obs}}{dt_0} \frac{dx_{phot}}{dt_{obs}}, \tag{7.9}$$

where x_{phot} symbolizes the notional position of a photon of propagating starlight. Dropping the $\pm$, supposing that $dx_{phot}/dt_{obs} = c$ (that is, the observatory measures light speed c), and using (7.8), we get

$$u = c\sqrt{1 - v_d^2/c^2} = \sqrt{c^2 - v_d^2} \ . \tag{7.10}$$

This agrees with the (higher-order) invariant neo-Hertzian result (4.9). This $u = c/\gamma$ value agrees with the light speed that would be measured by the observatory if it used a CT clock (compensated for earth's orbital motion) instead of the proper-time clock with which it measures light speed as c.

How does it come about that we made no use of the type of calculation embodied in Eqs. (7.7)-(7.10) in our earlier discussion of the mechanical equations of motion? Is there some difference? Indeed, there is a profound difference. In the case of stellar aberration our "observation platform" (the earth) is in non-inertial motion at instantaneous "detector" speed $v = v_d$, so that a correction factor γ^{-1} is needed to restore inertiality to the reference viewpoint. (Without non-inertiality of detector motion stellar aberration *does not exist* as an observable phenomenon, as we noted before.) Whereas in the mechanical case no such correction is needed: The observation platform can be considered to be in a permanently inertial state of motion, hence nominally "stationary," and the only (relative) motion is that of the various particles described.

The formal agreement of (7.10) with our previous neo-Hertzian result is perhaps at first surprising, but ultimately satisfying, in that the unravelling process whereby we undo the clock-compensation procedure just brings us back to where we stood originally in formulating *invariantly* the physics of the one-body (one detector) problem. The collective-time formulation accords with the proper-time one, invariant under the neo-Galilean transformation [(3.22), (3.23)], although now a different sort of invariance is involved — that of t_0 under the form of the Galilean transformation (to be discussed in Chapter 8, Section 3) that is expressed in terms of CT.

The principal advantage of introducing collective time is in treating not the one-body problem but the many-body problem, where the sort of glorified "frame time" it represents comes into its own as a simplifying descriptor. (It will be noted that the simplifying gain of integrability in Chapter 8 comes without loss of *formal* invariance under physical inertial transformations ... although the physical meaning of the invariance is altered because of the difference in operational definitions of "time.") In this book I shall not attempt to prove this many-body pudding, but take it as more or less obvious: It seems clear, or at least highly likely, that to replace an invariant proper-time descriptor that is non-integrable (path dependent) with a collective-time descriptor that is integrable (exact differential) opens the way to a consistent — and probably simplest possible — description of the configuration and interactions of many bodies. Once the effects of interaction on motion have been studied in the context of this simplified *integrable* formulation, a particle-by-particle translation back to results more directly observable in terms of natural (proper time) timekeeping can be accomplished for each particle independently, if desired, by the means (7.7) illustrated above for the single "detector" particle.

Of course, if we could accompany each particle of a many-body assemblage physically with its own *compensated* clock — as we do very readily in thought — then we could directly observe and prove or disprove the validity of the collective time approach to ensemble description. But that is generally out of the question in practical situations, as are most direct forms of experimental testing. I fear, then, that this is just one more theoretical incubus to be piled upon the back of the future ... and duly ignored. However, if any of the crucial experiments suggested in this book is ever done — *e.g.*, a VLBI test of stellar aberration — with an outcome in agreement with the present unorthodox predictions, there is a good chance that ignoration of the remainder of the present ideas will cease to be a scientifically attractive option. Therefore I am sustained by a hope for the future of physics unwarranted by the record of the past half century.

7.7 The light clock in orbit

It is of interest to find out what Hertzian electromagnetism, when formulated in terms of collective time t_0, predicts about an experiment to measure light speed in orbit. In such an experiment the measuring apparatus is at rest in an orbiting inertial system. What is the nature of the necessary apparatus? Let us simplify (perhaps over-simplify) by supposing it to have the geometry of a "light clock." We consider this to consist of two mirrors separated by a rigid rod, with a light pulse moving back and forth between them, and a cesium clock to time the period of oscillation. We stipulate a particular model of this oscillation process; viz., whenever a pulse arrives at a mirror it is absorbed ("detected") and immediately re-emitted. Consider things from the viewpoint of an orbiting observer co-moving with the apparatus. The Hertzian wave equation (7.5) has the solution (7.6). In this solution we place $\bar{v}_d = 0$, since the light detector (mirror) is at rest with respect to the observer and apparatus as a whole. Consequently Eq. (7.6) unambiguously predicts the measured phase speed to be $|u| = c$. Since in these equations time is being parameterized by compensated (CT) clocks, Hertzian theory, with $t = t_0$ and the assumption of length invariance, predicts that the *compensated* clock will measure light speed c, and the uncompensated (proper time) clock must consequently measure $c' \neq c$ —contrary to SRT and to Einstein's second postulate of light-speed constancy. (Actually, $c' = \gamma c > c$, since the uncompensated clock runs slower than the compensated one, and length invariance is assumed.) This supports an impression of the crucial nature of such an experiment.

Let us check this by switching viewpoints to that of the earth-surface inertial observer, with respect to whom the apparatus moves with speed v. We continue to consider measured time to be collective time t_0, as in the governing wave equation (7.5). Note that since the earth-surface observer is at rest with respect to the Master Clock used in defining CT, there is from his standpoint no distinction between CT and his own inertial frame time. Carrying out an analysis similar to that for Einstein's train in Chapter 6, Section 2, we consider the light clock device to move to the right (parallel to the paths of the light pulses) at speed v and a right-going light pulse to have speed v_r. If t_0' is the CT propagation interval for the pulse to pass a distance L from the left mirror to the right one, then $v_r t_0' = L + v t_0'$, which yields $t_0' = L/(v_r - v)$. Similarly for the return pulse having light speed v_l we have a return CT $t_0'' = L/(v_l + v)$. The total period of light-pulse oscillation, as measured by this earth-based observer, is then

$$T = t_0' + t_0'' = L\left(\frac{1}{v_r - v} + \frac{1}{v_l + v}\right).$$

For Hertzian electromagnetism, which we have seen to be appropriate for use in conjunction with CT, we have from Eq. (2.24), as in the train problem, $v_r = c + v$ and $v_l = c - v$. Putting in these values, we find $T = 2L/c = T_0$, where T_0 is the period for zero relative speed. This T is actually the oscillation period measured on earth, but, since CT in the orbital system is compensated to count "time" precisely in step with the earth-surface clock, the CT clock in the orbiting apparatus measures the same period, $T = T_0$, and L is the same whether measured on earth or in orbit. Hence, when time is measured by means of a compensated (CT) clock, the apparatus is predicted to measure two-way average light speed $2L/T_0 = c$, in agreement with our finding above for what the co-moving (orbiting) observer measures. Since the orbiting uncompensated (proper time) clock runs slower than the orbiting compensated clock by a γ-factor, thus showing less elapsed time per oscillation period, the former measures light speed $c' = \gamma c > c$, as previously deduced.

The problem of making a firm prediction of the outcome of the light clock experiment is complicated not only by questions of physical occurrence or non-occurrence of the Lorentz contraction, but by the necessity to employ a particular physical model of the light oscillation process. We assumed successive absorptions and re-emissions, so that there are two "detections" per cycle; but that model might be physically wrong. A different model pictures the light clock cavity as a "giant atom." In this case we might alternatively think of it as a sort of mode-locked Sagnac circuit. Previously, we pictured the muon and cesium atom as having a certain "sympathy" or similitude, such that their notional internal oscillations (controlling muon half-life) stayed in step regardless of the environment they might share. Here, we extend this model to the cesium atom and the light clock. If such a model or analogy is valid, there are no localizing absorptions at the mirrors, but instead the photon, when specularly reflected, stays in a quantum pure state, and its notional "oscillations" presumably bear a fixed ratio to the oscillations of the cesium atom. The result, for the experiment, would be as for the mode-locked Sagnac experiment— so the result predicted above on the absorption-reemission model would not be observed. The preceding Hertzian analysis would be inapplicable, since the assumed "detector motion" would not take place, inasmuch as the localizing "detection" events would not occur. Instead, under this "mode-locked" condition, I should expect the Einstein proper-time clock to measure c, since the mode-locked Sagnac apparatus behaves as if at rest, and under such circumstances of no relative motion it is the proper-time clock that measures c. One wonders if some such mode-locking may have affected the Michelson-Morley observations.

Is there any way such an (hypothesized) "apparatus effect" could be defeated, so that the "true" Hertzian effect of motion would be revealed (light speed c measurement by the CT compensated clock)? Here one might be further guided by the Sagnac analogy: Mode-locking was defeated in that case by vibrating or dithering the mirrors. Something of the sort might be tried here. I can only speculate. In a pinch, one could in principle design a modified apparatus—a light oscillator that was not a resonator or pure-state cavity, but that deliberately forced successive light beam absorptions and re-emissions. (That is, each mirror would be replaced by a separate absorber and emitter, with circuitry to pulse the emitter with minimum delay whenever an absorption occurred.) When such an apparatus was placed in orbit, one would be assured of dealing with true cyclic absorptions ("detections"). Einstein would make no distinction and would predict both for this and for the mode-locked apparatus that light speed would be measured (using proper time) as c. Whereas I would expect (for the proper-time clock) a difference—c for the mode-locked apparatus and $c' > c$ for the other. At least, this would not surprise me.

In the foregoing speculations I have addressed only motional effects, not gravitational ones. In the latter area there is no guidance from Sagnac experience, and one is flying blind. It is my guess that we possess no direct empirical knowledge as to whether there exists a mode-locking effect that could defeat gravitational changes of the running rate of a light clock. If there is no such effect, then gravity presents a surer test of my CT notions than does high-speed motion. Be that as it may, it is clear that an interesting, and rather broad, area of future research is suggested.

I could certainly be wrong about this, but I am going out on a limb in this book to predict that a *suitable apparatus* for average two-way light-speed measurement, when placed in orbit, will measure c with a CT clock (of the sort employed by the GPS) and $c' > c$ with a proper-time clock. This is the Hertzian prediction, as discussed above, in direct conflict with Einstein's second postulate. By "suitable apparatus" I mean one in which provisions are made to ensure its true oscillatory operation whereby genuine (localizing) absorptions of the photon occur during each cycle. That is, if such a phenomenon as mode-locking is found to occur in a moving light clock, then experimental measures are to be taken as needed to defeat it.

There remains the question, which of our two types of "time" (CT or proper time) is more useful for physics—which gives a simpler accounting of nature's doings when used in dynamical or field equations ("laws of nature") to describe particle interactions? It seems to me entirely too early to answer this question with final assurance. It cannot be answered by appeal to "principles," only to facts. Both types of time have their uses. We know that proper time is the measure of our localized personal experience of "time flow," whereas CT may be taken as the measure of our public experience of "now." But that is not the point in question. The point concerns dynamical description. My interim answer, in the perennial absence of full facts or divine revelation, rests on analogy: Thermodynamics would be a very complex subject, beautiful only in the eyes of long-indoctrinated specialists, if it were based on a definition of temperature T as what thermometers read. It becomes simple and elegant only when T is defined through Platonic idealization. As time-flow gauges, you and I and our buddy the muon resemble thermometers: we respond to our environment, but our aging in whatever environment need not be an accurate measure of the idealized time flow useful for simplest mechanical and electromagnetic description. A more trustworthy answer should emerge in due course, once physics is moved off its modern-day Ptolemaic dead center and re-established as an empirical science.

7.8 Two more forms of the relativity principle (distinguishable by a crucial experiment)

Allow me to expatiate briefly on two possible forms that Relativity Principle I (Section 7.2) might take, differing in their definition of the time parameter:

> *Candidate 1.* The laws of nature are the same in all inertial systems, when all clocks are allowed to run naturally without human intervention.

> *Candidate 2.* The laws of nature are the same in all inertial systems, when all clocks are compensated to run (display CT) in step with a Master Clock at rest in an inertial system, regardless of the choice of that fiducial system.

The first form is a rephrasing of the one discussed before (Relativity Principle I). The second, though unfamiliar, is physically as plausible, operationally as meaningful, and in better accord with the presently proposed alternative paradigm. The hidden sub-context is that Candidate 1, as employed by Einstein in his theory, tacitly assumes time dilation to be an appearance maintained symmetrically between any two inertial systems, whereas Candidate 2 treats it as a real, asymmetrical physical effect on clock rates. In Candidate 2 the choice of rest system of the Master Clock can alter the absolute rate of "measured" time flow—but this does not affect the "laws of nature" because its effects can be compensated by a units adjustment, as discussed in Section 7.3. (*Cf.* our consistency theorem of Section 7.2.)

Crucial experiment. These two alternatives submit to a decisive experiment, which is not simply a *Gedanken* experiment, but one that could be done practically and cheaply (as fundamental physics experiments are priced these days). Let a "suitable apparatus" for measuring light speed, together with a "dual-function" atomic clock (*i.e.*, one employing two counters, one set to measure the uncompensated proper-time second, the other compensated, in the manner of the GPS, to measure, when in orbit, the earth-surface Master Clock second) be placed in earth satellite orbit. Let light speed measurements be made in orbit with both clocks of this dual-purpose device, using the same apparatus and procedures. One of these measurements will yield the speed c, the other not-c. Candidate 1, in the SRT sub-context (including Lorentz contraction), predicts that the uncompensated clock will measure c. Candidate 2, together with our postulate of length invariance, predicts that the compensated clock will measure c. These predictions entail conflicting assumptions about space transformations; namely, in the first case occurrence of the Lorentz contraction (at least as a symmetrical appearance—let's not dig too deeply!), in the second non-occurrence of any length changes (real or apparent) in the measuring apparatus. Such an experiment should cast light on (a) the two candidate forms of relativity principle, (b) the validity or invalidity of Einstein's second postulate (light-speed constancy referred to proper-time measurement), and/or (c) the occurrence or non-occurrence of the Lorentz contraction.

If, on performance of the experiment, the uncompensated clock is observed to measure c, this will support the validity of SRT in respect to (a) Candidate 1 for the relativity principle (or, more generally, Relativity Principle II), (b) Einstein's second postulate, (c) occurrence of the Lorentz contraction. If the compensated clock measures c, this will directly refute SRT and will support (a) Candidate 2 for the relativity principle, (b) non-constancy of light speed, (c) length invariance. I predict the latter, unorthodox, result. All forms of the relativity principle discussed herein clearly embody one expression or another of the "relativity spirit." To decide among them is a matter for empiricism.

7.9 Kinematics for uncompensated clocks

In this section (only) I revert briefly to proper time and uncompensated clocks. Back in the bad old days of classical mechanics, when Newton's absolute time prevailed, little was heard about "kinematics," or even about "Galilean transformations." This was because the corresponding description was a fairly trivial affair. Inertial systems were important to Newton's laws, of course, but how things looked from the viewpoint of one or another different one of these was no big deal, because they were all equivalent, by a Newtonian relativity principle. Everything was simple: the frame-time differential dt was both invariant and exact (integrable). The transfer between systems of mensuration standard units for calibrating time and space gave no conceptual trouble, because such transfer did not affect the metric properties of the units. This is still true, as far as I know, for space units (my opinion and postulate) ... but the new element for the description of nature claimed in this book, and empirically well supported, is that time units are affected (in an explicitly known, asymmetrical way) by any such transfer. There are two methods of dealing with this: (a) Einstein's way, which is to employ a natural (proper) time as whatever uncompensated clocks *measure*, and (b) the way advocated in this book, which is to *compensate* natural clocks so as to remove from their measurements the effects of all energy state changes—and to define as a *collective time* that which is measured by a set of such compensated clocks disseminated throughout space. There should be room for both kinds of "time," proper and collective, in the physics of the future, since both submit to operational definition and both (I suggest) have their distinct merits and uses.

Let us briefly review the most elementary aspects of the kinematics associated with uncompensated or proper time and invariant length—just as a reminder of the flavor of the subject. If one uses naturally-running uncompensated clocks (which might be called "Einstein clocks") then inertial transformations, in view of length invariance, are described by a "neo-Galilean transformation" of the form

$$\vec{r}' = \vec{r} - \vec{v}t \qquad t' = \frac{t}{\gamma} \qquad\qquad (7.11a)$$

[*cf.* Eqs. (3.20)-(3.23)] with inverse

$$\vec{r} = \vec{r}' + \vec{v}t = \vec{r}' - \vec{v}'t' \qquad t = \gamma t', \qquad\qquad (7.11b)$$

which shows length invariance and velocity non-reciprocity, $\vec{v}' = -\vec{v}\gamma$. Here the primed inertial system S' moves with speed $|\vec{v}| < c$ with respect to the unprimed system S, as measured by instruments at rest in S. Whereas S moves with respect to S' with a velocity γ-times greater than $\vec{v}$, as measured by instruments at rest in S', hence possibly superluminally. Note a formal discrepancy: If primes are put on all symbols of the space transformation in (7.11a), with double primes interpreted as unprimes, a true relationship [the spacelike transformation of (7.11b)] is obtained; but if the same is done to the time transformation in (7.11a) we get $t = t'/\gamma'$, which would match the time transformation in (7.11b) only if $\gamma' = 1/\gamma$. The latter contradicts $\gamma' = \gamma$, which is supported by

$$\gamma' = \frac{1}{\sqrt{1 - \dfrac{v'^2}{c'^2}}} = \frac{1}{\sqrt{1 - \dfrac{(v\gamma)^2}{(c\gamma)^2}}} = \frac{1}{\sqrt{1 - \dfrac{v^2}{c^2}}} = \gamma. \qquad (7.12)$$

(Here $c' = c\gamma$ echoes our finding in Section 7.7.) This discrepancy, expressing a loss of formal symmetry, arises from our insistence (based on GPS evidence) that the asymmetry of timekeeping between primed and unprimed systems is objectively real. It gives rise to proper-time velocity non-reciprocity and seems to be another indicator supporting our asserted difference between time and space—in this case a basic difference of their symmetry properties (space invariant, hence symmetrical between inertial systems; proper time objectively asymmetrical) ... another nail in the coffin of *spacetime*.

The frame-time parameters t, t' appearing in (7.11a,b) represent a "natural time," measured by ordinary Einstein clocks that are not compensated in any way, but are allowed to run at their unaltered (proper-time) rate in whatever environment they may be placed. This would refer to atomic clocks, muons, biological aging, *etc*. In this book my thesis is that for purposes of analysis it is far more convenient to introduce an "unnatural time" (collective time), such that all clocks—compensated to make it true—run at the same rate regardless of environment.

In this connection presumably an "absolute time" would be a collective time referred to a Master Clock at rest in some *physically preferred*, unique, "fundamental" system. If the preference were related to the hypothesized rest state of a physical "ether" responsible for light propagation, this would correspond to Maxwell's original ideas or to Lorentz's ether theory—the latter being historically the source of the Lorentz transformations. But Lorentz's ether is a physical one that mechanically—by exerting stress—affects observable lengths of moving extended structures. His theory would appear to differ predictively from SRT in that a reduced speed of ether wind relieves stress and allows structural expansion, while increased speed produces structural contraction; whereas in SRT only contractions are normally considered to occur.

However, SRT squirms aside from any attempt to pin it down unambiguously on such issues. Thus Kelly[2.13] attributes to Einstein "in 1919" the following remark: " ... a rigid disc in rotation (produced by casting) must explode as a consequence of the inverse changes in length, if one attempts to put it at rest." By the "casting" remark he presumably means that the disc is rotated while in a liquid state and allowed to solidify (freeze without dimensional changes) during rotation. Evidently the "inverse changes" he has in mind amount to a Lorentz expansion. Kelly remarks that, "It is interesting that Einstein states here that the Lorentz contraction is a real observable phenomenon." Perhaps Einstein has discovered a new explosive. Unfortunately, to get the kinematic transformation equations to support the explosion as a "real observable phenomenon," γ and $1/\gamma$ would seemingly have to play reciprocal roles, as for time in Eq. (7.11a,b) ... and that would contradict the Einsteinian symmetry of his version ("*Candidate 1*") of the relativity principle, reflected in the symmetry of the Lorentz transformation.

Evidently we encounter here more amphibiousness between appearance and reality, symmetry and asymmetry (*cf.* Chap. 6, Section 3). It would seem that Einstein sat amphibiously on both sides of this barbed-wire fence. You will find some of his followers on each side, with not a few able to follow the master in feeling comfortable on both sides. Modern physical theorists, after enough years of homeopathic adjustment to absurdity as a way of life, have apparently lost all capacity for wonder at absurdity.

Lorentz's conception of an ether possessed of a determinate state of motion is as foreign to my ideas as it is to Einstein's. By contrast, the present conception of a collective time places the Master Clock at rest in an arbitrary inertial system—chosen not on the basis of a physically dictated "preference," but at the analyst's whim.

7.10 The need for more facts

It has been pointed out to me that mutual contradictions lurk among the numerous "principles" and "postulates" enunciated in this chapter governing natural or proper-time timekeeping. The fact is that at this stage of our knowledge far too little is known empirically—about how uncompensated clocks run naturally under various conditions of environment and transport—to allow proposing a trustworthy set of axioms. Einstein's heaven-storming attempt and crashing failure to achieve this should serve as a chastening lesson illustrating the folly of trying to do all physics with no better tool than the mind. The axiomatic "method," when employed in theoretical physics, amounts to shooting in the dark and hoping to be lucky. If physics is a serious enterprise it deserves better. Since the time of Galileo, physicists have known, and have lately forgotten, a faster converging process. The mind is an impressive instrument ... but what it is best at impressing is itself.

It is readily apparent that various *kinds* of clocks behave in quite different ways. For instance gravity-actuated devices, such as pendulum clocks, water clocks, and hourglasses, when placed in a gravity-free environment (*e.g.*, a satellite in orbit) cease to "tell time" at all. There is no scheme of compensation that can correct this. For them, time "exists" only in non-inertial systems! If, as we have done here, we focus attention on atomic timekeeping, this has the virtue of simplifying our operational considerations. But, led on by this simplicity, we are in no small danger of fooling ourselves about the extent of our knowledge.

For example, can the reader point to any experiment with atomic clocks that unambiguously verifies the state function assumption (our Postulate I)? Or that even verifies its reversibility corollary, to the effect that a clock transported from state of motion A to state B, and then back to A, will resume its original running rate when at rest in A? I mentioned the deficiencies of Hafele-Keating in this regard: One is forced to suppose that the experimenters would have bothered to check the running rates of returned clocks after round-the-world transport, and further to suppose that if they did check they would have faithfully reported any anomalies, and to deduce from this that because they reported no anomalies none arose ... and to suppose all these things about people known[2.13] to have no other purpose in undertaking the experiment than to "confirm" the foreknown truth revealed by a sacred theory. What kind of reputation ensues in the real world from reporting anomalies at variance with a sacred theory? History answers this by pointing to the case of D. C. Miller, a respected experimentalist of unassailable integrity and competence, who reported anomalies in his repetition of the Michelson-Morley experiment and was promptly declared incompetent by theorists, and is so judged almost unanimously[2.13] by physicists to this day.

It is possible that further investigation will show not only logical contradictions in the proper timekeeping scheme outlined in this chapter, but also how to remove such contradictions by dropping or modifying one or more of the premises herein stated. As the likeliest example, the state function assumption might fail. Although this question is basic to the physics, and can be settled only by a program of combined empirical and rational investigation, its resolution is not essential to the main topic of this chapter, which is collective time. Regardless of how clocks behave empirically, as long as they behave *reproducibly*, it requires only the most minor act of faith to suppose that in all significant cases *GPS-type engineering can cope with the challenge of finding compensation methods to permit a consistent CT definition.* This does not in the slightest lessen the priority of experiments needed to expose the facts of timekeeping with different clock types under a full range of environmental conditions and transport protocols. We need to know those facts just to be able to call ourselves physicists. Meanwhile, however, the contemporary incompleteness of knowledge in that department need not inhibit development of alternative physical theory (mechanical, electromagnetic, and quantum) based on time parameterization by CT.

7.11 Chapter summary

The GPS engineers recognized motional and gravitational effects as among the annoying irrelevancies to consistent global timekeeping that had to be eliminated by suitable clock-rate compensations of definitional character. By such compensations they were able to establish rate synchronism with an inertial Master Clock among a collection of clocks in essentially arbitrary states of motion and gravity environments. This provides us with a model and operational definition of a concept of collective time (CT) t_0. If mathematicians demand an existence proof for CT, I offer the GPS. If they claim that the concept contains logical contradictions, again I offer the GPS and ask if facts contradict themselves. Indeed, the only contradiction is of the doctrines of a different paradigm or establishment of faith.

The rate at which CT "flows," for a single stage of clock compensation involving $\gamma(v^2)$, depends on choice of the Master Clock's inertial rest system. This appears to make the concept incompatible with a relativity principle. However, I have given several arguments supporting a contrary view, such as that the "laws of nature" referred to in Form I of the relativity principle are independent of the absolute flow rate attributed to "time." Numerical flow-rate changes can be compensated by adjustment of the arbitrary time unit (units being distinct from laws of nature). Alternatively, a second stage of clock compensation can be applied to all clocks to impose any preferred nominal flow rate. I conclude that CT is compatible with a valid form of the relativity principle; namely, "*Candidate 2*" proposed herein (Section 7.9). This differs from Einstein's "*Candidate 1*," which I have shown by multiple arguments in Chapter 6 to assert a timekeeping symmetry among inertial systems incompatible with GPS evidence. For Einstein's proper-time clocks can show a real (asymmetrical) time dilation only with the help of a different form of "compensation" — *viz.*, a real (asymmetrical) Lorentz contraction — to allow light speed to be measured as c in all environments. As far as I know, nobody has ever proposed that the Lorentz contraction can be asymmetrical. Yet, every time the CERN muon topic is raised, everybody accepts that time dilation can be asymmetrical.

To a limited extent it is possible to work out a kinematics based on time as told by uncompensated clocks, at least in respect to one-body motions, based on a "neo-Galilean transformation." But it seems impossible that this could ever be extended in any simple way to describe a many-body dynamics, in view of the inexact nature of $d\tau$. For simplicity in treatment of the many-body problem there seems to be no substitute for a CT formulation employing the integrable (exact) differential dt_0. This allows t_0 to be used optionally (a) as a coordinate in a Euclidean (3+1)-space (affine) geometrical representation, (b) as a parameter in a 3-space representation $\vec{r}(t_0) = (x(t_0), y(t_0), z(t_0))$ of Lagrangian character, or (c) to define a phase-space representation allowing formal recovery of classical canonical mechanics with all its impressive group properties. In terms of CT, kinematics trivializes to the true Galilean form, mechanics trivializes to the Newtonian (or Lagrangian or Hamilton-Jacobian) form with a "relativistic" mass increase, and field theory trivializes to the Hertzian form — all with t_0 playing the role of formal analog of the classical absolute time.

This does *not* mean that classical disabilities, such as the inability of Hertz's theory to describe stellar aberration, are inherited. On the contrary, when the implications of clock compensation entailed in the definition of t_0 are duly taken into account, consistency with neo-Hertzian invariant theory (based on proper time) and with observation is obtained. Evidently there is no reason CT t_0 should not be introduced into "non-relativistic" quantum mechanics as a painless way of "relativizing" the formalism (probably with great benefit to treatment of the quantum many-body problem), with none of the complications of a covariant formulation. Thus all fields of physics appear receptive to parallel reform. The (politically obligatory) claim that Einstein's theories are the only ones capable of covering the known range of empirical physical knowledge is laughable. To Einstein I concede unchallenged reign only in gravity's domain, wherein his esteemed theory rests firmly on a rotten foundation (SRT).

Analytic simplifications of the sort offered by the "old physics" here advocated (CT analogs of Galilean kinematics, Newtonian mechanics, Hertzian field theory) will be of interest to *real physicists*, currently trapped *incommunicado* in the "new physics," only if the experiments here recommended yield the unexpected results I have predicted. To review them:

(1) A VLBI measurement of stellar aberration angle, for which SRT predicts a specific non-zero departure from Bradley aberration at second order (in the ratio of earth's orbital speed to light speed), and I predict no departure until third order.

(2) A measurement of light speed by "suitable apparatus" physically placed in orbit, using a dual-function atomic clock reading both proper time and CT, for which SRT predicts that the proper-time clock will measure c and I predict that the CT clock will measure c.

The first of these is simple, conceptually unambiguous, and clearly crucial. About the second I am much less certain … it seems simple but is subject to multiply branching pathways of analysis that make prediction hazardous. There is potentially enough "physics" of the unwanted apparatus-effect sort to make this experiment less clearly crucial. But I suggest it should be done, because a lot will be learned regardless of the outcome.

References for Chapter 7

[7.1] P. W. Bridgman, *The Logic of Modern Physics* (MacMillan, New York, 1927).

[7.2] G. H. Hardy, *A Mathematician's Apology* (Cambridge, 1940, 1946, 1976, 1992, *etc.*)

[7.3] F. Hoyle and J. V. Narlikar, *Action at a Distance in Physics and Cosmology* (Freeman, San Francisco, 1974).

[7.4] A. K. T. Assis, *Relational Mechanics* (*Apeiron*, Montreal, 1999).

[7.5] P. Graneau and N. Graneau, *In the Grip of the Distant Universe: The Science of Inertia* (World Scientific, Singapore, 2006).

Well, I'll be—space isn't curved after all!

> —B. Thaves, *Frank and Ernest*, 11 January, 1990.
> (Comment of a hatching chick)

Chapter 8

Linkages of Time, Energy, Geometry

8.1 Connection of time and action (and the effect of gravity)

This final chapter is primarily an amplification of selected earlier material. From Eq. (3.3a), (6.13), (7.1a), (7.7), *etc.* we see that the factor f, by which a compensation is effected that nullifies a clock's running rate decrease due to work being done on it to change its motional state, is

$$f = \frac{d\tau}{dt} = \frac{1}{\gamma}. \tag{8.1}$$

We know that this γ-factor is related to the *total mechanical (kinetic plus rest) energy* E_{mech} of a moving clock of rest-mass m_0 by

$$E_{mech} = \frac{m_0 c^2}{\sqrt{1-(v/c)^2}} = m_0 c^2 \gamma ; \tag{8.2}$$

hence we can write γ as

$$\gamma = \frac{E_{mech}}{m_0 c^2}. \tag{8.3}$$

This result is pregnant with occasions for "vivid contemplation" (to borrow a phrase from the late Senator Arthur Vandenburg). It displays γ as a dimensionless ratio of total mechanical energy to rest energy. This is interesting in itself, since γ entered our theorizing as a dimensionless ratio of pure time quantities, Eq. (3.3) or

(8.1). Now suddenly we see it morphing into a dimensionless ratio of pure energy quantities. Is this really the same γ? Of course. Then we may be pardoned for inferring some deep-lying physical connection between energy and time.

This may be obvious to modern quantum sophisticates, who recall from the $\Delta v \Delta t \sim 1$ of wave theory, with $E = h\nu$, that $\Delta E \Delta t \sim h$, a relationship of "complementarity" between time and energy. But here we get to a more explicit relationship of actual equality between timelike and energylike ratios,

$$\frac{dt}{d\tau} = \gamma = \frac{E_{mech}}{m_0 c^2}. \tag{8.4}$$

The existence of a relationship between time and energy has not always been recognized by physicists. For instance Heinrich Hertz, who has been aptly described as "the most philosophically profound of the great nineteenth century physicists," had no hesitation in declaring flatly[8.1] that "There exists no connection between mass and time alone." So, if we now assert such a connection, we have either progressed beyond earlier perceptions or have deluded ourselves. In any case we shall here further explore the perception or delusion.

First comes a question: Why should our concern stop with total *mechanical* energy? A more interesting quantity to consider would surely include potential as well as mechanical energy. However, in this context the algebraic sign of potential energy is negotiable. Our attention being directed to something of the nature of "total energy," we are led to propose the introduction of a generalized form of γ designated γ_{tot} and [by analogy with (8.4)] defined by

$$\gamma \rightarrow \frac{E_{tot}}{m_0 c^2} \rightarrow \gamma_{tot} = \frac{E_{mech} + |V|}{m_0 c^2} = \frac{m_0 c^2 \gamma - m_0 \Phi}{m_0 c^2} = \gamma - \frac{\Phi}{c^2}. \tag{8.5}$$

Here $m_0 \Phi$ is the potential energy of mass m_0 (for instance an orbiting clock in the earth's gravity potential of the GPS system). The numerator $E_{mech} + |V|$ is equivalent (apart from an additive constant) to a quantity a, the integrand of *action*, discussed below. It resembles total energy $E_{tot} = E_{mech} + V$ except for the sign of $V = m_0 \Phi$. That sign is chosen to ensure that γ_{tot} is a definitely positive quantity, Φ being assumed in all cases to be negative. Thus, the gravitational potential at distance r from a point mass M (the earth) in Newton's mechanics is customarily taken to be the negative quantity

$$\Phi = -\frac{GM}{r}, \tag{8.6}$$

the gauge being chosen to cause the potential to vanish at infinity. Observe that for $r \to 0$ this quantity can approach $-\infty$, which means that E_{tot} can do the same. This disqualifies total energy to represent γ_{tot}, since it is desirable that any physically acceptable form of γ satisfy $1 \le \gamma < \infty$. Hence the need in (8.5) to change the sign of $V = m_0\Phi$.

This choice of the minus sign for Φ in (8.5) is highly significant. It means that we are replacing "total energy" E_{tot} with a modification of energy having the character of what in integral form is traditionally called "action." To see this, we note that action A in mechanics is formally defined[5.3] as the definite time integral (using collective time t_0 as parameter)

$$A = \int_{t_1}^{t_2} a \, dt_0 \tag{8.7a}$$

of

$$a \equiv \sum p\dot{q} = L + H = T - V + H = E_{mech} - V + const, \tag{8.7b}$$

where $E_{mech} = T + m_0c^2$, T is kinetic and V is potential energy, $L = T - V$ is the Lagrangian, and we assume *conservation* (constancy) of total energy H, the Hamiltonian. Thus, with disregard of additive constants, the action differential a and total energy E_{tot} differ only in the sign of V or Φ. The potential sign choice is necessary as discussed above, and also as required to agree with observation. Since H is constant, it is apparent from (8.7b) that "a" in (8.7a) can be interpreted either as $\sum p\dot{q}$ or as the Lagrangian L. (Although the Lagrangian is normally considered to have no physical interpretation, we may infer from the above its interchangeability with "action" in conservative problems, for which $H = const.$)

The form (8.6) is unsatisfactory in general because it makes the numerical value of γ_{tot} depend on an arbitrary gauge choice. It is convenient for present purposes to eliminate the gravity potential effect on earth surface clocks by choosing instead the gauge

$$\Phi = -\frac{GM}{r} + \frac{GM}{R}, \tag{8.8}$$

such that $\Phi = \Phi_{surf} = 0$ at $r = R = $ earth's radius. In this case we can define a potential *difference* between orbiting and surface clocks,

$$\Delta\Phi \equiv \Phi_{orb} - \Phi_{surf} = \Phi_{orb} = -\frac{GM}{r_{orb}} + \frac{GM}{R}. \tag{8.9}$$

Since numerically $\Phi_{orb} = \Delta\Phi$, it follows that $\Phi = \Phi_{orb}$ in (8.5) can be replaced formally by $\Delta\Phi$:

$$\gamma_{tot} = \gamma - \frac{\Delta\Phi}{c^2}, \tag{8.10}$$

a form preferable to (8.5) because it shows no gauge dependence. Thus we focus attention on changes of energy or action state. Let this new γ_{tot}, defined by (8.10), be substituted for γ in the definition (8.1) of f:

$$\gamma \to \gamma_{tot} \qquad f \to 1/\gamma_{tot}. \tag{8.11}$$

If N_0 is the number of atomic clock oscillations per earth-surface second, defined as what we choose to mean by the second measured by a Master Clock located there, then the correction factor to be applied to that number for clocks in orbit is $f_{orb} = 1/\gamma_{tot}$, and the corrected number of oscillations for orbiting clocks is

$$N_0' = f_{orb}N_0 = \frac{N_0}{\gamma_{tot}} = \frac{N_0}{\gamma - \Delta\Phi/c^2}. \tag{8.12}$$

In passing we observe that our gauge choice $\Phi_{surf} = 0$ is needed for consistency with the formal requirement that $f_{surf} = 1$. This may be seen from the fact that, if we were to apply (8.5) at the earth's surface, with $v_d = 0$,

$$f_{surf} = \frac{1}{\gamma_{tot\,surf}} = \frac{1}{\gamma_{surf} - \Phi_{surf}/c^2} = \frac{1}{1 - \Phi_{surf}/c^2}, \tag{8.13}$$

with any other gauge than $\Phi_{surf} = 0$, this would imply $f_{surf} \neq 1$. (At the earth's surface we obtain consistency by considering a Master Clock of the GPS system at rest, $v_d = 0$, $\gamma_{surf} = 1$, and $\Phi_{surf} = 0$, so that $f_{surf} = 1$ and $f_{surf}N_0 = N_0$, as must be true by definition.)

For comparison with GPS observations of orbiting clocks we have $\gamma = \left(1 - v_{orb}^2/c^2\right)^{-\frac{1}{2}} \approx 1 + v_{orb}^2/2c^2 + \cdots$ and from (8.12), to first order in c^{-2},

$$f_{orb} \approx \frac{1}{1 + v_{orb}^2/2c^2 - \Delta\Phi/c^2 + \cdots} \approx 1 - \frac{v_{orb}^2}{2c^2} + \frac{\Delta\Phi}{c^2} + \cdots. \tag{8.14}$$

Observe that the algebraic sign of $\Delta\Phi$ here is correct [for $r_{orb} > R$ Eq. (8.9) shows $\Delta\Phi > 0$, which in (8.14) rightly indicates gravitational and motional corrections of opposite signs]. It is my understanding that the expression on the right in (8.14) is indeed the correction factor (for "relativistic" motional effects and for New-tonian gravity, leaving out any other corrections of less funda-mental interest) applied by the GPS engineers in getting all their clocks successfully to run at the same rate regardless of state of motion. The gravitational correction is observed to be in the op-posite direction from the motional one. (I gather that in the GPS geometry the gravity correction is about six times larger in mag-nitude). The upshot is that the connection between action and time explored in this section seems to check against GPS observa-tions. This confirms our use of the negative sign for Φ in (8.5) or for $\Delta\Phi$ in (8.10).

I confess to considerable curiosity about the physics underly-ing the role of what I have chosen to identify as "action" in the above analysis. Allow me to speculate briefly. We are concerned with the transfer of atomic clocks from one state of motion and gravity potential to another. Such clocks are quantum systems in essentially pure states, and the transfer processes are strictly con-servative; that is, the system Hamiltonian remains constant throughout. (The proof of this conservation is recovery of clock rate upon recovery of initial energy state.) Because of exactness of the differential dt_0 of CT in (8.7a), the path by which clock trans-fer takes place is immaterial—the results are path independent. This means that the transfer path integral with respect to collec-tive time of some descriptor of the system must have a stationary property. That is, its variation, subject to constancy of the Hamil-tonian (and allowing variations of the time endpoints on paths of varying length as needed to maintain such constancy), must van-ish. In effect all physical paths are equivalent, so their variation is nil.

It is well known[5.3] that the classical descriptor having the in-tegral property just described is a, the system's *action* density, not its total energy. This stationary property (for variations of the stated type) is known as the Principle of Least Action. It is for this reason, I opine, that the substitution of a for total energy in (8.5) makes sense. If I am correct about this, then other applications should arise in quantum mechanics. For example, *when any quan-tum pure state system moves across gravitational equipotentials, that motion should be governed by the Principle of Least Action.*

It may be added that the foregoing results, based on action, by no means represent a unique alternative, nor do they conform at higher orders to traditional approaches. GPS observational confirmation extends only to the first order in c^{-2}, so entirely different treatments, differing at higher order, are possible. For example Hatch in his very interesting book[6.6] treats the gravitational effect on clock rates by means of its own separate gamma factor,

$$\gamma_g = \frac{T_0}{T_g} = \frac{1}{\sqrt{1 + \dfrac{2\Phi}{c^2}}}, \tag{8.15}$$

T_0 being the time of a clock outside the gravity field (where $\Phi = 0$) and T_g the time of a clock in that field at the position where Φ is evaluated. His results, employing a *product* of gravitational and motional gamma factors to describe the combined effect, are in first-order agreement with those given here [right-hand side of (8.14)]. (Similarly, Van Flandern[8.2] employs the product approach, but uses the notation σ for Hatch's γ_g^{-1}.) Only at (currently untestable) higher orders in c^{-2} would distinctions emerge.

    Addendum. After this was written, Ronald Hatch kindly called my attention to an important paper by Charles M. Hill ["Timekeeping and the Speed of Light—New Insights from Pulsar Observations," *Galilean Electrodynamics* 6, No. 1, 3-10 (1995)], which casts an interesting sidelight on the approach I have advocated. Hill espouses the idea of a Newtonian time, albeit in the context of a natural "substrate" as preferred reference system. He concludes that "it is reasonable to adopt the pulsar second as a tentative Newtonian time standard until a better approximation is found." He believes that the effects of an hypothesized objective variability of light speed with environment are just such as to cancel the objective variability of Einstein's proper time (which he calls "einstime"); so that constancy of the measured c-value in all environments results. If he is right about this, I am wrong in my prediction that, for light-speed measurements in orbit, compensated rather than uncompensated clocks will be needed to yield the c result. (I see no reason to accept his hypothesis of exact compensatory light speed variation with environment. It conflicts with both Maxwellian and Hertzian theory.) He also gives for what I am calling the *f*-factor of clock rate compensation yet a different expression from anybody else's. All these expressions differ at second order in c^{-2}. It may be useful to tabulate them here:

My way, based on action change:

$$f = \frac{1}{\gamma - \Delta\Phi/c^2} = 1 - \frac{v^2}{2c^2} + \frac{\Delta\Phi}{c^2} + \frac{1}{c^4}\left(-\frac{v^4}{8} + (\Delta\Phi)^2 - v^2\Delta\Phi\right) + O(c^{-6}).$$

Hatch-Van Flandern, based on GRT??:

$$f = \sqrt{\left(1 - \frac{v^2}{c^2}\right)\left(1 + \frac{2\Delta\Phi}{c^2}\right)} = 1 - \frac{v^2}{2c^2} + \frac{\Delta\Phi}{c^2}$$

$$+ \frac{1}{c^4}\left(-\frac{v^4}{8} - \frac{(\Delta\Phi)^2}{2} - \frac{v^2\Delta\Phi}{2}\right) + O(c^{-6}).$$

Hill, based on ?:

$$f = \sqrt{1 - \frac{v^2}{c^2} + \frac{2\Delta\Phi}{c^2}} = 1 - \frac{v^2}{2c^2} + \frac{\Delta\Phi}{c^2}$$

$$+ \frac{1}{c^4}\left(-\frac{v^4}{8} - \frac{(\Delta\Phi)^2}{2} + \frac{v^2\Delta\Phi}{2}\right) + O(c^{-6}).$$

I have taken the liberty here of substituting my $\Delta\Phi$ for the Φ used by these authors. It will be observed that all three expressions agree at the observable first order in c^{-2}, but disagree at the second order. If that order ever becomes observable, some decision may emerge. I assume that General Relativity Theory (GRT) might enter up to n other candidates in the contest, given n choices of the metric function from a non-denumerable infinitude of possibilities.

8.2 An effect of gravity on mass in equations of motion?

Now arises an interesting question—one of those choice points of theory open to debate. It will be observed that the basic mechanical equation of motion, (7.4), contains explicitly a γ-factor. The question is, should this γ be interpreted at face value, as in traditional physics, or should it be interpreted as γ_{tot}, in accordance with prescription (8.10)? The latter interpretation breaks new ground, physically, as it proposes an entirely general effect of potential energy change upon particle effective "mass" (without mentioning gravity or invoking an equivalence principle). Such effects have been suggested at various times and in various contexts, but the matter seems to remain rightly somewhat controversial. I do not know of empirical evidence that decides, and can think of theoretical arguments either way. It would be interesting to compare the predictions of Einstein's "curved spacetime" analysis with the simple Euclidean-space consequences of combining (7.4) and (8.10). That is,

$$\vec{F}_0 = \frac{d}{dt_0} m_0 \left(\gamma_0 - \frac{\Delta\Phi}{c^2} \right) \vec{v}_0. \qquad (8.16)$$

This entails the assumption that ancillary relations such as (7.1)-(7.3) involving γ also obey the replacement rule (8.10). The comparison with general relativity theory, however, could hardly allow a definitive choice, since GRT—most grandly conceived—is not an actual theory but a program for a class of theories—inasmuch as the "metric" is up for grabs. It is an adjustable *function* ... which goes Ptolemy's elastic set of adjustable parameters one better. In other words, GRT is super-Ptolemaic; it is not a falsifiable theory until completed by the addition of a falsifiable element (typically, the Schwarzschild metric). This topic lies beyond my own competence or interests, but it may interest others who follow, if any. It would be amusing, and highly worthwhile, if the present Euclidean-style analysis should prove competent to guide future space missions. (Though of course it could never improve on the *accuracy* of GRT methods, just as Keplerian astronomy cannot improve on the accuracy of a Ptolemaic astronomy favored with enough adjustable constants.)

I am assuming that back at home base enough information exists to be able to calculate both γ and $\Delta\Phi$ with "sufficient" accuracy to determine the *f*-factor at all points on the trajectory of a practical space voyage. If not, the voyage should probably not include human passengers. Given the necessary information for clock compensation everywhere, one can conceptualize "all space" as filled with clocks in various states of motion, in the manner of Einstein's "clock gas," except all clocks have their running rates compensated so as to tell a common CT t_0 in step with some inertial Master Clock. These compensated clocks need not themselves be in inertial states of motion but can be moving arbitrarily.

The advantage for astronautics? I should think it would greatly simplify all dynamical calculations to be able to assume a general "now" (with instant gravity action and rigorous obedience of Newton's third law), and to steer the space vehicle on the basis of compensated clock readings and a quasi-Newtonian analysis of trajectories [*i.e.*, Newtonian with allowance for mass increase according to (7.4) or (8.16)], using Newton's law of instantaneous gravitation. Compare having to use tensor symbols to describe motions governed by natural clock (proper time) readings with curved space geodesic dynamics, geometrodynamics, or whatever ... and maybe, given the trend, by next year doing it in Hilbert space with an infinite set of tensor indices.

8.3 Kinematics for compensated clocks

In Section 7.7 a brief introduction was given to the neo-Galilean transformations, Eq. (7.11), which assume uncompensated clocks. Let us address now the more interesting case of what happens when one uses only compensated clocks. In that case one moves in a single bound all the way back in history to Galilean kinematics. Physics has had an exciting century consorting with sophisticated mathematically-painted whores, and here we are back in mother's womb. That is, the "inertial" transformations are formally the old-fashioned Galilean ones,

$$\vec{r}' = \vec{r} - \vec{v}_0 t_0 \qquad t_0' = t_0 . \tag{8.17}$$

The indicated invariance of the measured time parameter is strictly an artefact of our clock-compensation scheme. That makes it no less "real," since invariance is a purely mathematical property that makes no distinction between fictions and realities. Recall that the subscript "0" on t_0 refers to the collective time (CT) readings of atomic clocks in arbitrary gravitational environments and states of motion, contrived to agree with an inertial Master Clock by resetting their internal counters in such a way that if N_0 is the number of atomic oscillations per proper-time (or Master Clock) second, and N_0' is the corresponding number for the CT "second," then, per (8.12),

$$N_0' = f_0 N_0 = \frac{N_0}{\gamma_0 - \Delta\Phi/c^2} . \tag{8.18}$$

The method thus specified ensures that all CT clocks, when placed in their intended states of motion, run at (approximately) the same rate as the Master Clock. In order that they be in phase synchronism (that is, truly "in step"), so that they always read the same elapsed "time," special methods are needed. If light signals are used, they must be described by Hertzian electromagnetism, discussed in Section 7.3, rather than by Maxwellian electromagnetism. However, such signals (though essential in practice) may not be necessary in principle. Rate synchronism can in principle be achieved, without distant signalling, by clock transport described on the basis of detailed foreknowledge of the environmental conditions along the path of transport. Thus, one can (*Gedankenweise*) consider the fundamental equation linking elapsed proper time to "frame" time, Eq. (3.8a), to be generalized in the present case to

$$\tau = \int \sqrt{1 - v_0^2/c^2}\, dt_0 = \int \frac{dt_0}{\gamma_0} \rightarrow \int f_0 dt_0 = \int \frac{dt_0}{\gamma_0 - \Delta\Phi/c^2}\, ; \qquad (8.19)$$

so that actual times (clock phases) can in principle be deduced by integration of combined motional and gravitational histories along the path of transport.

There is, of course, a large gap between such "principle" and practicality, since at every point of an accelerated journey the $N_0' = f_0 N_0$ value used in clock resetting would in general change and require continual resetting. Calculation would be neither easy nor reliable. If the clock arrived at a more or less stable state of motion (unaccelerated—ideally Galilean inertial motion), characterized by a fixed N_0' value, fixed v^2, fixed $\Delta\Phi = 0$, etc., a single synchronization operation by radio would suffice to allow it to tell collective time, the $t_0 = t_0'$ of (8.17). In practice, however, as is true of the GPS, space-traveling clocks would need continual monitoring for correction of drifts, inaccuracies, and environmental changes.

It is apparent from (8.17) that the velocity composition law for compensated clock measurements is the Galilean addition law,

$$\vec{v}'' = \vec{v}' + \vec{v}. \qquad (8.20)$$

Proof: Let a Master Clock, at rest in inertial system S_0, define universal time t_0 everywhere. Let Particle 1 move in S_0 with velocity $\vec{v}' = d\vec{r}_1/dt_0$ and let Particle 2 move in S_0 with velocity $\vec{v}'' = d\vec{r}_2/dt_0$. The position vectors are considered to be parameterized by t_0, $\vec{r}_1 = \vec{r}_1(t_0)$, $\vec{r}_2 = \vec{r}_2(t_0)$. The velocity $\vec{v}$ of Particle 2 relative to Particle 1 is the rate at which their separation $\vec{r}_2 - \vec{r}_1$ changes with t_0, as measured by any observer employing CT t_0. That is, $\vec{v} = d(\vec{r}_2 - \vec{r}_1)/dt_0 = \vec{v}'' - \vec{v}'$, *q.e.d.*

The result applies to all observers, since both length and time are formally invariant under general coordinate transformations corresponding to arbitrary motions of reference systems—

because the compensations of clock rates apply to all particles of such systems, and all refer to the fiducial state of motion of some inertial Master Clock. Thus general invariance does not make the claim of general covariance, to rid physics of the special status of inertial systems. For that, an entirely different theme is needed. Presumably, the *physical cause* of inertiality, involving Machian instant action of distant matter of the universe, needs to be brought in (Assis,[7.4] Graneau and Graneau[7.5]). Inertia is thus viewed as a problem of physical causation, not of formal ballroom dancing with geometry.

Note that the separation $\vec{r}_2 - \vec{r}_1$ (an invariant length, the same for all observers) is an instantaneous one measured at any instant of CT. This is exactly as in Newton's physics and is unintelligible in Einstein's physics—an entirely different paradigm. Similarly, velocity reciprocity holds between inertial systems,

$$\vec{v}' = -\vec{v}, \tag{8.21}$$

or between any two observers in relative motion at any instant of CT. [This is the special case of (8.20) in which the observer co-moves with Particle 2, $\vec{v}'' = 0$.]

At once it will be objected that (8.20) violates the basic rule that particle velocity relative to the Master Clock's inertial system must not exceed c. But that rule is already enforced automatically by the clock compensation scheme—is, so to speak, built into it. For each clock (notionally associated with each particle in the universe) is compensated [*cf.* Eq. (8.18)] by multiplying the Master Clock's oscillation number N_0 by a function f of γ, a quantity that becomes imaginary for $v > c$ and infinite for $v = c$. If we exclude infinities and imaginaries as non-physical, we see that $v < c$, where v is velocity relative to the Master Clock, is physically required for all compensated clocks, whatever their locations and states of motion. The use of compensated clocks is preconditional to the meaningfulness of Eq. (8.17) or (8.20). It must therefore be recognized that superluminal particle or clock motions in S_0, the rest system of the Master Clock, are *not* allowed by (8.20), despite superficial appearances. The nonlinearities associated with high-speed motions are implicit (present but hidden) in the formally "linear" relations (8.17).

It is convenient to speak, as I do here, of CT speeds and CT velocities as those measured by compensated clocks. Although all particle CT speeds are limited to c in S_0 (*i.e.*, relative to the Master Clock), they are not so limited with respect to each other. It is possible for *relative* CT speeds to approach as an upper bound $c + (-c) = 2c$ when velocities are oppositely directed. This is a perfectly good relative velocity, since it could be directly determined as a rate of change of separation distance by either of two near-c observers measuring their separation by means of compensated clocks, or by any other observers not participating in their relative motion. That value ($2c$) is the maximum measurable relative CT speed, since all observers and particles are limited to c in CT speed with respect to the Master Clock's state of motion. In any composition of CT speeds each summand must individually obey the c speed limit with respect to the Master Clock (to which all clocks are notionally connected by transport). Note that c itself is defined in terms of CT. Moreover, c ceases to play the central group-parameter and metric-defining role that it does in Einstein's theory. But its demotion as a physical descriptor (of light speed) has already been foreshadowed in Hertzian theory.

It is clear that CT introduction offers extensive possibilities for formally simplifying physical description. It reconciles persistent classical demands for instant action-at-a-distance description with practical operational definability. Unfortunately, the information requirements for such an operational definition are much greater than for the time concepts employed in traditional or in Einsteinian physics. One really needs to know many details of the environment in which a clock is to be immersed before one can conceptualize or perform the compensation operations needed to enable it to tell CT. In principle, however, I feel that this is not qualitatively different, for example, from the environmental information demands regarding temperature variations made upon accurate metrology with material length standards.

A more serious difficulty is that the *analysis* needed to translate the simple answers obtained in the Euclidean-Galilean-Newtonian CT format into answers useful in the real world of environmental influences (analogous to the "spacetime curvature" of Einstein's world) may prove not much easier than dealing with the current four-index tensors. We have already indicated the concealed nonlinearity of the problem. In principle a mechanical problem can be set up, for example, in the classical canonical formulation, and solved to yield a formal answer, just as in the nineteenth century, but in terms of CT t_0 rather than Newtonian time. In such a solution each particle for which we desire knowledge of its actual proper-time "aging" must have its clock "decompensated" by inverting the procedures of its compensation—doubtless a tedious and inelegant computational problem. What is gained on the swings may thus be lost (computationally) on the round-abouts. I have no slick program to offer for this translation. If I promised you a rose garden, I retract. One could work out simple demonstration problems, as for textbooks. But a general formalism must remain for the future—its development contingent on a change in the currently prevailing climate of refusal to consider alternatives to SRT.

Reverting to the idea of a dual-function clock, one could imagine every notional particle in the universe provided with one of these, comprising an Einstein clock that tells "natural," uncompensated, or proper time, and a compensated clock that tells CT. The natural clock informs the particle how to age, when to get hungry, *etc.*, whereas the compensated clock tells it how to navigate by simplest equations of motion, how to interact with other particles by the simplest laws, *etc.* For operational definition, the same notional bunch of cesium atoms could serve for both clocks, the difference between them being merely the choice of number of oscillations assigned to the "second" ($N'_0 = f_0 N_0$ *vs.* N_0). Therefore in thought, if each particle could carry with it a dual-function clock, no computations would be needed, and the many-body problem's *post facto* solution would consist in a log or record for all particles of where each particle was at all collective times and how old it was when it was there.

With reference to time measurement by compensated clocks, Eq. (8.17) indicates that physical *inertial motions* are linear and uniform, without rotation or Thomas precession. (Clock-rate fiddling doesn't change any of that from the original Newtonian case.) Rigid bodies have six degrees of freedom, as of yore. Extended structures exhibit no anomalies, mechanical systems possess well-defined and unique centers of gravity, *etc.* The geometry of phonograph turntables remains Euclidean, whether or not they are set into rotation. Alas, I am about out of discussion topics, at just the point where SRT begins spinning its endlessly fascinating fairy tales. The reason I do not find much to say about the Galilean transformations is that it has all been said.

Since I have alluded to the Thomas precession, perhaps I should conclude with a word on that. Supposedly an early "confirmation" of SRT (through observation of an inferred effect of *electron spin*), the Thomas precession is a house of straw that needs very little huffing and puffing to blow it down. From our previous results, or from *Heretical Verities*,[2.11] we can express the invariant momentum of a particle as

$$\vec{p} = m_0 \left(d\vec{r}/d\tau \right) = m_0 \vec{v}\gamma = m_0 \vec{v}/\sqrt{1 - v^2/c^2} \; .$$

Solved for $\vec{v}$, this yields $\vec{v} = \vec{p}/\sqrt{m_0^2 + p^2/c^2}$. Using this to eliminate $v^2 = \vec{v} \cdot \vec{v}$ in favor of p^2 from the expression for $\gamma = 1/\sqrt{1 - v^2/c^2}$, we find $\gamma = \sqrt{1 + (p/m_0 c)^2}$. Putting this into the classical one-body Hamiltonian H, equal to the total energy, we get

$$H - V_0 = E_{mech} = m_0 c^2 \gamma = m_0 c^2 \sqrt{1 + (p/m_0 c)^2} = \sqrt{(m_0 c^2)^2 + p^2 c^2} \; ,$$

[with the help of (8.2)] where V_0 is the scalar potential energy. Finally, squaring both sides of this equation,

$$(H - V_0)^2 = (m_0 c^2)^2 + p_x^2 c^2 + p_y^2 c^2 + p_z^2 c^2 \; ,$$

and performing the Dirac linearization (linear operator equivalent of the square root of the quadratic form—which is a perfectly good, albeit unconventional, classical 4×4 matrix operator), we get formally the Dirac Hamiltonian of quantum mechanics, which needs only the provision of a wave-function operand {with formal Correspondence, *cf.* Eq. (5.19b), whereby we make replacements

$$\vec{p} \rightarrow \vec{p} + (e/c)\vec{A} \rightarrow (\hbar/i)\vec{\nabla} + (e/c)\vec{A} \text{ and } H \rightarrow -(\hbar/i)\partial/\partial t_0 \; ,$$

these quantum limits being easily remembered from the classical Hamilton-Jacobi relations $\vec{p} = \vec{\nabla}S$ and $H = -\partial S/\partial t_0$, with Hamilton's principal function S replaced formally, $S \rightarrow \hbar/i$, and the operators allowed to act on an operand (four-component wave function) to their right[2.11]} to yield the Dirac electron theory. The latter fully "explains" electron spin without need for help from either Thomas or SRT. Note that SRT, although it agrees with it, was not *used* in obtaining this "relativistic" result. (The only input to our argument borrowed from Einstein is Eq. (3.1) or (3.3), which depends on the invariance of the timelike interval $d\tau$ along the particle trajectory, *not* on the spacelike $d\sigma$.) The magic "Thomas" factor of ½ appears automatically in Dirac's theory—a god from another machine.

8.4 Velocity composition: More than you wanted to know

Having glimpsed the heaven of CT simplicity, let us for our sins briefly return to the purgatory of proper-time complexity. Concerning the composition of velocities, SRT's rule is that, given objects 0,1,2 at rest in the correspondingly numbered inertial systems, such that object i has speed v_{ij} relative to object j (*i.e.*, as measured by instruments at rest in system j), the law of one-dimensional speed composition is explicitly nonlinear,

$$v_{10} = \frac{v_{20} + v_{12}}{1 + v_{20}v_{12}/c^2}.$$

Thus if system 0 is our laboratory the speed of object 1 measured there differs at $O(c^{-2})$ from the simple sum of its speed relative to 2 and 2's speed relative to the lab. The reason for this exotic form is that clocks in the different systems run at different natural (uncompensated) rates, resulting in speed "measurements" that do not jibe, but can be adjusted by the formula. However, to be perverse, consider the difference $v_{21} - v_{10}$. In SRT we have $v_{ij} = -v_{ji}$, so this difference is

$$v_{21} - v_{01} = v_{10} - v_{12} = \frac{v_{20} + v_{12}}{1 + v_{20}v_{12}/c^2} - v_{12}$$

or

$$v_{21} - v_{01} = v_{20} - (v_{20} + v_{12})v_{20}v_{12}/c^2 + O(c^{-4}).$$

Since the three inertial systems are equivalent, we can identify any of them as the "laboratory." Suppose we take system 1, where object 1 is at rest, as the laboratory. Then this relation means that when lab observer 1 measures separately the collinear speeds of objects 2 and 0 and takes the difference, he gets a relative speed between objects 2 and 0, as measured by 1, which we might designate v_{201}. We have to introduce a new notation here for what observer 1 measures, since the result is not equal to the relative speed v_{20} of 2 as measured by 0, but instead differs from that at $O(c^{-2})$ because of the different clock rates in systems 1 and 0. It is to be emphasized that this new quantity $v_{201} \equiv v_{21} - v_{01} \neq v_{20}$ is an "observable." It is defined by *measuring* in system 1 the two individual speeds v_{21} and v_{01} and *calculating* the difference. So, it is an operationally definable observable for which there is no place in SRT as normally developed. (We mentioned its counterpart above in reference to two oppositely directed photons with relative CT speed $2c$.) Physically, it represents the time rate of change of the separation distance between 2 and 0 in 1's inertial system, as measured by 1. If that does not qualify as 1's idea of "relative speed of 2 and 0," what does?

Why bring this up? Simply because observer 1 might want (or even urgently need, if he is a controller responsible for two satellites in near collision) to know the relative speed of 0 and 2, and he has no other way (intrinsic to his own system) to quantify the idea of such a relative speed except to *difference* the two numbers he has obtained by separate measurements with his own clocks and yardsticks. One of the premises of SRT is that each inertial system is equivalent to any other and just as good, if not better. So, observer 1 is fully entitled to his own opinion about relative speeds, based on his own measurements. He need not kow-tow to other observers, nor inquire how their clocks are running. He paints his world picture, they paint theirs. I emphasize again: v_{21} and v_{01} are observer 1's *own* measurements, numbers he reads off his co-moving instruments. For him, these two numbers separately quantify the speeds of objects 2 and 1 relative to his own inertial system. There is no relevance of clock *rate* differences—they do not enter observer 1's problem. Yet SRT not only de-emphasizes this simple speed-measurement differencing but has no natural place in its notation or formalism for the concept "relative speed of 2 and 0 as measured by 1." Ordinarily, the idea is not recognized as qualified for admission to the Einstein-Minkowski "world" because, as we have seen, it can lead (in terms of the frame time of inertial system 1) to $v_{201} = c - (-c) = 2c$.

In this connection, it is amusing to note, a recent issue of *Science News* (Vol. 169, No. 20, page 319) informs us that

> Researcher David N. Spergel agrees that general relativity requires that no object move through space faster than light. He adds, however, 'General relativity also predicts that space itself can expand. ... We can actually point to distant galaxies, on opposite sides of the sky, that are moving apart from each other at faster than the speed of light.'

If, indeed, "space itself can expand," this means it possesses the physical property of elasticity. We know all about elasticity from SRT—it is that property of bodies which enables them to rotate. (Remember? If a body lacked elasticity it would be Born rigid, and if it were Born rigid it would be logically forbidden to rotate—*cf.* the Herglotz-Noether theorem.) So, general relativity theory (GRT) is another one that uses purely mathematical axioms to "predict" a physical property, is it? What meat has such theory fed on? You don't get physical garbage out of any mathematical theory without putting it in at the start. Actually, the "physics" deduced in such cases is invariably a form of emergency surgery to stop arterial bleeding of the logic of the theory. According to Einstein, GRT was solidly built upon SRT. SRT was built upon c as a limiting speed in nature. And GRT, without contradicting SRT, "predicts"—in flat contradiction of SRT—that something called "space" long ago exhibited a physical property of spectacular inflationary elasticity but, in agreement with SRT, no longer does so today because if it did we would measure speeds greater than c in our lab. However, we can look at galaxies in opposite directions today and see this elasticity at work— while, according to SRT's "worldline" concept, long ago and today and the distant future are all the same, any distinction being physically meaningless (because observers in different states of motion disagree about them). And if long ago and today are the same, this means that near and far are the same ("Far or forgot to me is near ... When me they fly, I am the wings"), because of spacetime symmetry, so separations of objects in our lab and of distant galaxies are the same—and lab space is elastic, after all, like the critical sense of the relativist. If you buy all or any of that, there is a bridge in Brooklyn I'd like to sell you ... and a tonic that long ago would have grown hair on a billiard ball, though not today—except that today and long ago are the same, so it might be worth an open-minded trial at your risk, $179.95 the bottle plus postage. (Yes, dear relativist, the foregoing is a wickedly unprincipled misrepresentation of your oh-so-confirmed theories ... *Peccavi*.)

SRT was notable for eradicating metaphysics by eradicating the "ether." That is to say, it got credit for banishing airy fancies and establishing the realities of "measurement" as the sound workaday foundation underlying all physics. Einstein then turned around in the 1920's and revived the ether, thereby getting credit for banishing plodding realities in favor of the products of unfettered mathematical imagination *via* non-Euclidean geometry. With metaphysics thus put firmly back in the saddle, "space" became the new ether, which we have seen to be endowed with the physical property of elasticity. This opened the way for singu-

larities (breakdown points of equations, where they yield operationally meaningless infinite answers), in the form of Black Holes,
to acquire physical properties, such as temperature, and to grow
hair, as on a billiard ball. It does not matter where this will end,
and we need not ask, since the beginning is already a total disas-
ter, and totality surely sufficeth.

In stark contrast, the CT formulation eradicates all complexi-
ties from the outset by providing all observers with a common
"time" parameter t_0. The velocity composition law is the elementary vector addition law of (8.20). With the use of CT, the whole
rigmarole above can be boiled down to $\vec{v}_{10} = \vec{v}_{20} + \vec{v}_{12}$, or equiva-
lently $\vec{v}_{20} = \vec{v}_{10} - \vec{v}_{12} = \vec{v}_{21} - \vec{v}_{01}$, with no second-order "corrections." All observers agree on everything and there is no going-
outside the formalism, no need for new notations, *etc.*, to treat
any observer's viewpoint. It is thus a notable relief to have arrived finally at a simpler theory in which the testimonies of all
observers agree, regardless of their motions or measurement protocols—and there is only one law of velocity composition (the
Galilean one)—because all observers use CT time. Such agreement, I emphasize, has never been possible in SRT because of the
lack of a shared time parameter.

To summarize thus far: In SRT, when all measurements are
made *in a single inertial frame*, velocities compose by the classical
law of vector addition, not by any higher-order "composition"
law. This does not contradict anything in the canons of SRT, nor
indicate self-inconsistency. But it may call into question the attributing of observed relative speeds of galaxies exceeding c to a
GRT elastic property of "space"—since relative speeds up to
$v_{201} = 2c$ are perfectly well accommodated by SRT for measurements made in the earth's single inertial frame. And it leaves a
black mark on the aesthetics of SRT, in that it reveals the need for
two separate formal laws of velocity composition, the choice depending on whether the measurements being described take
place in a single reference system (linear law) or in multiple systems (non-linear law). This is neither neat nor pretty. But,
then ... neither is it neat nor pretty that in SRT there are two
separate laws of length transformation—longitudinal and transverse—as measured in a single system. It shows the power of
mass persuasion that generations of physics aesthetes have found
beauty in such ugliness.

Although I have no right to assume the reader to be a glutton for the kind of punishment involved in the study of proper-time kinematics, I have here made the beginnings of a start on that dismal topic and might as well go a bit farther down this same barren road. We shall stick to our physical premises of length invariance and asymmetrical natural timekeeping of clocks subject to energy state changes (*i.e.*, undergoing objectively real, nonreciprocal rate changes). The formal consequences can be worked out, as follows, by turning cranks already provided.

Reverting to the two particles of Section 8.3 in arbitrary states of motion, with a Master Clock at rest in inertial system S_0 reading time t_0, we recall that the velocity of Particle 1 as measured in S_0 is $\vec{v}' = d\vec{r}_1/dt_0$ and that of Particle 2 is $\vec{v}'' = d\vec{r}_2/dt_0$. Let τ_1 be the proper time (*i.e.*, that told by the co-moving natural or uncompensated "Einstein clock") of Particle 1, and τ_2 that of Particle 2. From our basic relation (3.3) we have

$$d\tau_1 = \frac{dt_0}{\gamma'}, \qquad \gamma' = \frac{1}{\sqrt{1 - v'^2/c^2}} \qquad (8.22a)$$

and

$$d\tau_2 = \frac{dt_0}{\gamma''}, \qquad \gamma'' = \frac{1}{\sqrt{1 - v''^2/c^2}}. \qquad (8.22b)$$

An observer moving with Particle 1 and using that particle's proper-time clock to measure its velocity with respect to S_0 will evaluate that frame-relative "proper velocity" as

$$\vec{V}' = \frac{d\vec{r}_1}{d\tau_1} = \frac{dt_0}{d\tau_1} \cdot \frac{d\vec{r}_1}{dt_0} = \gamma'\vec{v}', \qquad (8.23a)$$

and similarly for Particle 2,

$$\vec{V}'' = \frac{d\vec{r}_2}{d\tau_2} = \frac{dt_0}{d\tau_2} \cdot \frac{d\vec{r}_2}{dt_0} = \gamma''\vec{v}''. \qquad (8.23b)$$

Such proper velocities, as previously mentioned, can be superluminal. From (8.22) we note the sometimes useful relation

$$dt_0 = \gamma'd\tau_1 = \gamma''d\tau_2 \;\rightarrow\; \frac{d\tau_2}{d\tau_1} = \frac{\gamma'}{\gamma''}. \qquad (8.24)$$

The proper velocity of Particle 2 relative to Particle 1, as measured by the Einstein clock co-moving with the latter, takes the form

$$\vec{V}_{12} = \frac{d(\vec{r}_2 - \vec{r}_1)}{d\tau_1} = \frac{d\vec{r}_2}{d\tau_1} - \frac{d\vec{r}_1}{d\tau_1} = \frac{d\tau_2}{d\tau_1}\frac{d\vec{r}_2}{d\tau_2} - \frac{d\vec{r}_1}{d\tau_1} = \frac{\gamma'}{\gamma''}\vec{V}'' - \vec{V}', \qquad (8.25a)$$

which with (8.23) can also be written as

$$\vec{V}_{12} = \frac{\gamma'}{\gamma''}\gamma''\vec{v}'' - \gamma'\vec{v}' = \gamma'(\vec{v}'' - \vec{v}') = \gamma'\vec{v}, \qquad (8.25b)$$

where use has been made of (8.20). Similarly, the proper velocity of Particle 1 relative to Particle 2, as measured by an uncompensated clock co-moving with 2, is

$$\vec{V}_{21} = \frac{d(\vec{r}_1 - \vec{r}_2)}{d\tau_2} = \frac{d\tau_1}{d\tau_2}\frac{d\vec{r}_1}{d\tau_1} - \frac{d\vec{r}_2}{d\tau_2} = \frac{\gamma''}{\gamma'}\vec{V}' - \vec{V}'', \qquad (8.26a)$$

which can alternatively be expressed as

$$\vec{V}_{21} = \frac{\gamma''}{\gamma'}\gamma'\vec{v}' - \gamma''\vec{v}'' = \gamma''(\vec{v}' - \vec{v}'') = -\gamma''\vec{v}. \qquad (8.26b)$$

Note, on comparing (8.25b) and (8.26b) that the relative proper velocities, as measured by each of the two particles, are quasi-reciprocal, in the sense of being oppositely directed (opposite signs) along the same line in space, but with different magnitudes, owing to the objectively different running rates of their proper-time clocks. Considering the magnitudes of those two expressions and taking the ratio, we get with (8.24)

$$\frac{V_{12}}{V_{21}} = \frac{|\gamma'\vec{v}|}{|-\gamma''\vec{v}|} = \frac{\gamma'}{\gamma''} = \frac{d\tau_2}{d\tau_1} = \frac{(clock\ rate)_1}{(clock\ rate)_2}, \qquad (8.27)$$

which confirms that the measured relative proper speeds are in the inverse ratio of rates of the clocks doing the measuring.

Hardy souls who crave more thrashings-about in the tar pits of proper-time parameterization can find a record[2.11] of my own lengthy and futile struggles. For those of more normal psychology, it should suffice to have discovered an alternative to all this—a formal escape back to classical kinematics—a way of rendering the many-body problem tractable by systematically setting clocks so that path integrals become path independent and closed path integrals become zero. Any schoolchild can see the advantage, and any GPS engineer can and does effectuate the method.

8.5 The many-body problem: γ as integrating factor

Although the reader is probably fed-up with the subject, it may be instructive to reflect a bit further upon the factor γ that keeps cropping up ubiquitously in all discussions of accurate timekeeping or energetic motion. Formally, what it does in its *Ur*-form, by Eq. (3.3a), is to link the differentials of proper time and frame time, $dt = \gamma d\tau$. It will do no harm to re-emphasize that in physics differentials may have two distinct and important properties, invariance and exactness. In Newtonian physics, where there is an absolute time, dt possesses both of these desirable properties, so they are not much noticed. (You never know your luck till you lose it.) But with the advent of SRT, proper time split off conceptually from frame time. In the aftermath, proper time took with it the invariance property, but lost exactness, whereas frame time kept the exactness, but seemed to lose invariance. That being the case, looking at the above expression linking the two types of differential, we see that the role played by "γ" is (mathematically) that of a formal *integrating factor*, which converts the inexact differential $d\tau$ into an exact differential dt. (For coordinates quite generally form exact differentials—it would be awkward to label axes with path-dependent quantities!—whereas we know the line integrals of $d\tau$ to be path-dependent ... *cf.* the net aging difference of the famous twins.)

Why should it be important to deal with exact differentials and coordinate quantities, rather than with invariant proper times? One good reason: somewhere in the background of all physics lurks the many-body problem. You do not hear much about this in SRT. In fact, it has become a sort of buried secret—a topic seldom brought to the frontal lobes of the brain. Textbooks of SRT do not have chapters headed "The Many-Body Problem." Did you ever wonder why? Of course not, but, while I am forcing you to listen to so much else you don't want to hear, I'll address the question. The reason is that the theory, by its basic terms of reference, is not well adapted to handling or even formulating the many-body problem. Consider a single particle moving arbitrarily in an inertial frame. Its description *via* proper time in Minkowski space is beautiful and elegant. But how are we to connect this to the Lorentz transformation, which is fundamental to all SRT description? By attaching an inertial frame to the particle that instantaneously co-moves with it, so that it is momentarily *at rest* (proper time and frame time being the same for a particle at rest). In this frame we know the physics—it is pre-relativistic. Well and good ... so far. But inevitably the headstrong, pesky particle changes its state of motion, and then what do we do? Obviously, we have no recourse but to attach another co-moving frame and discard the old one. And barely have we succeeded in that arduous task than we have to do it all over again ... In short, we end up busier than the proverbial one-armed paper-hanger, just keeping up with our frame-attaching obligations to one miserable particle.

Next consider a collection of many differently-moving particles. You see? The mind reels. For now things get much worse. There is no single inertial frame in which, even for an instant, even as few as two particles can co-exist at mutual rest. For two or more particles there seldom exists even fleetingly a *co-moving inertial frame* to "attach." So, the whole clever dodge of attaching a co-moving frame collapses, *spurlos versenkt*. Not exactly a selling point for any theory.

What's to be done? Within SRT, nothing. To make progress we must stand outside SRT and recognize that we are dealing with apples and oranges. We are asking the oranges, the solipsistic particle proper times described by inexact differentials, to talk to the gregarious apples, the frame times characterized by exact differentials. And they simply will not converse. That's what all the "frame attachment" frustration is mutely trying to tell us. In order to establish communication between these different fruit species, we must exploit the integrating factor by applying it separately to each individual particle's "orange" proper-time differential—in order to convert it to its "apple" counterpart, a frame-time differential ... and of course, in order to encourage a consistent dialog, the *same inertial frame* must be chosen in all cases. This is precisely what is achieved by the γ^{-1}-factor (or f-factor) *compensation* operation employed in defining the collective time t_0 —the operation being performed separately for each of an assemblage of differently-moving clocks—clock-setting consistency being assured by referring all compensation operations to a single inertial (Master Clock) state of motion. By compensating the clocks co-moving with various particles, we are simply applying *integrating factors* to the corresponding particle proper times of our many-body problem. These integrating factors are so chosen as to convert the inexact differentials $d\tau_i$ of proper time (oranges) into exact differentials $dt_0 = \gamma_i d\tau_i$ of the coordinate time t_0 of a common inertial system (apples). Where gravity changes along trajectories, we may suppose that the integrating factor generalizes, $\gamma \to \gamma_{tot}$, where γ_{tot} is given by (8.10).

Once we get all these apples talking together on the same wavelength (to mix mixed metaphors), we have the basis for a serious attack on the many-body problem. In fact, its difficulty is reduced to that of a quasi-Newtonian many-body problem—which is well-known to be analytically very formidable, indeed, but at least straightforwardly computable. [Actually, because of mass increases at high speeds, Eq. (7.4) or (8.16), the problem is technically a bit more difficult than the Newtonian one, but not much. The great simplification over SRT is that the Euclidean geometry of a single inertial frame can be employed ... with no frenzy of "frame attachment."]

Why is it important thus to acquire (for the first time) a real grip on the "relativistic" many-body problem? Simply because of the central role played by that problem in all physics. Since GPS satellites do not much interact gravitationally, it is not important in the GPS context to be able to solve it. But in other contexts the subject comes to center stage. For instance, in quantum mechanics emergent properties such as "collective" actions have become associated already with the non-relativistic many-body problem. Biophysics cannot mature as a theoretical science without some conceptual grip on the many-body problem. The attempt to extend quantum mechanics to the nuclear domain meets the difficulty that simple-seeming entities such as the proton may well be very many-body assemblages of much smaller entities moving at relativistic speeds.

Instead of being able to meet such problems head-on, by applying a viable many-body mechanics of the high-speed regime, the absence of such (owing to the interminable, all-conquering, all-crippling SRT assault) has forced upon physicists dazzling exercises of imaginative virtuosity—creative fantasies such as second quantization and the endless ugly larding-over of point-particle mechanics with its conceptual antithesis, continuum field theory. The result, after many years of malpractice built upon malpractice, is a kludge that only a mother could love. And, boy, do those mothers love it. They love it so much that when empiricism produces evidence of a new phenomenon, "cold fusion," occurring in the many-body solid state regime (which happens, however, not to fit their many-body preconceptions), they have such abiding love for their monstrous but familiar theoretical contrivance that they vigorously suppress not only all publication but all experimentation in the new field, thereby causing it to wither in the bud.

Finally, it may be remarked that the advantages of a reversion to Newtonian methods have been only lightly touched on here. Perhaps most important is the restoration of the lovely nineteenth-century formalism of classical canonical mechanics— Hamilton-Jacobi theory and all that good stuff that led naturally to the elegant *non-relativistic* canonical formalism with its huge *invariance* group. It is likely that the modern spavined form of that theory, crippled to fit universal covariance, can now be junked. There used to be a truism that "progress" in theoretical physics could be measured by increases in the size of the invariance group recognized by physics, with the classical canonical group the all-time champion and high-water mark. This involved true invariance not only under "general" coordinate transformations, but under much fancier and more exciting scramblings of coordinates and momenta.

All that got quietly jettisoned (or blindfolded and made to walk the plank) when physics took its giant step backwards under the impact of the universal acceptance of spacetime symmetry. In the aftermath of that intellectual *tsunami*, mere covariance under general coordinate transformations came to be regarded as a tremendously big deal ... and forever after got taught to the young and gullible as the *ne plus ultra*. At least one perceptive student of the scientific scene[8.3] has observed that science does not progress in a linear fashion. But I daresay at few moments in its history has physics taken in a single bound so great a leap backwards, nor instantly and discontinuously placed itself farther from its previously recognized goals.

8.6 A fable

As a change of pace, let me offer a brief semi-philosophical excursion into the land of fable. Imagine a universe in which *all materials* display a piezoelectric effect (*i.e.*, all lengths are affected by electricity). And imagine that the physicists of this hypothetical universe unanimously support an inflexible "measurement" doctrine to the effect that lengths are *by definition* what their meter sticks measure. Thus the effect of electricity is not to perturb the accuracy of a meter stick but to alter distance (length) itself. (I know ... it is hard to imagine stupidity of this magnitude, but make a special effort.) They understand enough empirically about "laws of nature" to be aware of how to compensate the effects of electricity on meter sticks, but at the basic level of their physical conceptualization of "length" they treat that knowledge as irrelevant.

The laws of nature they have formulated are accurate, work well, and satisfy them completely, but are inordinately complicated because they have to incorporate the compensations (of electrical effects on "length") that they decline to apply directly to the meter sticks themselves. In other words, their high-level laws are bent all out of shape by the obligation to do low-level work. Since electrical environments may differ in different inertial systems, there can be no physically valid relativity principle in such a universe (governing the over-complicated, bent-out-of-shape, *form* of their "laws of nature"), because lengths, metric properties, and the metrics themselves, can differ among the different systems, as well as within a given system.

In this imaginary universe there is no "equivalence" among inertial systems and no "invariance" of the laws of nature. In each system valid laws can be formulated, but those laws are not formally invariant across systems because the metric properties vary with electrical conditions. Instead, with the help of multi-index tensor symbols, the laws are considered to possess properties of "covariance" that encourage these hypothetical physicists to *assume* a relativity principle premised on a communal metric, without ever seeking empirical evidence regarding what goes on in inertial systems and environments other than their own. That is, they reason circularly that a relativity principle *holds* because the idea is so beautiful and because it works (leads to no contradictions of experience) in their own system … and since it holds they need not go to the trouble and expense of putting observers into other systems to test it literally. (I know I have strained your credulity. Forgive me. I do hope it isn't fractured.)

Into this kingdom of the blind comes a wayfaring stranger whose open eyes perceive at once that the "laws of nature" (1) can be simplified, and (2) can be made the same in all inertial systems—*i.e.*, can be made rigorously invariant, so that a relativity principle truly holds—by the elementary expedient of re-conceptualizing "length," not as what meters sticks measure, but as a Platonic ideal approximated by what meter sticks measure in the limit of zero electrical perturbations. Although the stranger's message is obviously true, and would bring untold benefits by being heeded, I regret to report that in this particular hypothetical universe it is ignored, the physicists deny him publication, and he dies early of a broken heart. The physicists continue to thrive in their faith in a relativity principle, without stooping to test that principle by so much as a single experiment employing measurements made in different inertial systems under different environmental conditions ... and life and tenure go on as before, to the tune of blithe reassurances that their theory is the most extensively experimentally tested in the sidereal universe and in the history of animate thought.

Now, luckily, our actual universe is quite different. Our physicists are not stupid. They recognize that electrical effects on length should be compensated, and that it is "time," not length, that demands definition *via* an inflexible "measurement" doctrine. That is, time is *defined* as what clocks measure, with no compensation for the action of known laws of nature that dictate an effect of energy state changes (motional and gravitational) on clock running rates. Thus the effect of energy state changes is not to perturb the operation of a clock but *to alter time itself*. Now, to be sure, it is obvious, and hardly needs the ministrations of a wayfaring stranger to point out, that the actions of known environmental effects, if uncompensated, preclude the validity of a relativity principle ... for in different inertial systems different environmental conditions may prevail asymmetrically, with consequent non-invariance of metric properties, of metrics, and of the "laws of nature." Also it is obvious that the laws of nature (1) can be simplified, and (2) can be made the same in all inertial systems—*i.e.*, can be made invariant (with none of the artful-dodging of "covariance"), so that a relativity principle does honestly hold—by the simple expedient of re-conceptualizing "time," not as what clocks measure, but as a Platonic ideal approximated by what clocks measure in the limit of zero energetic perturbations. Since in this actual universe of ours, as noted, the physicists are very smart, it follows that they need only have the just-stated observations called to their attention and they will rush to correct their thinking and simplify their laws. They will fall all over themselves to accept such enlightenment for publication and to heed it. I warned that I was going to tell you a fable. This last bit about our "actual" universe is the fable.

Perhaps I should "document" my fable by furnishing evidence that it has a firm basis in the accepted beliefs of our era. Throughout my last three chapters I have argued for a conception of *time* as an idealized extrapolation from what clocks record, freed of environmental associations (*e.g.*, effects of motional and gravitational energy state). Just how far this lies from the current party line among physicists may be judged by a quotation from one of their gurus, Daniel Kleppner, an ex-MIT physics professor and director of the MIT-Harvard Center for Ultracold Atoms, who writes in one of his authoritative monthly "Reference Frame" articles in *Physics Today* (March, 2006, p. 11),

> At first sight the problem of gravitational potential appears to be yet one more mundane experimental factor that must be controlled to operate an atomic frequency standard, much like temperature, magnetic field, or laser intensity. However, there is a fundamental distinction: the effect of gravity is not to perturb the operation of a clock but to alter time itself.

There you have the MIT-Harvard Ultracold Truth. Oh, dear ... so the GPS engineers really goofed in not clearing with Physics Authority their misguided mundane impulse to control atomic clocks. In seeking to compensate the effects of gravity on such clocks they were tinkering unbeknownst with the very fabric of the universe, with time itself, and probably putting the times out of joint. Who knows what subtle and far-reaching damage these left-footed, pig-ignorant engineers have wrought already? In the lurid annals of heedless human malfeasance, global warming isn't a patch on it. Cry scandal, cry havoc! Cry *New York Times*, *Time* magazine! The Bird of Time has been winged. Maybe we aren't even in the twenty-first century. I'm wondering ... maybe it's a good thing that not all engineers go to MIT-Harvard. As for physicists ... well, to look on the bright side, maybe that's the best place for them—leaving Princeton aside, of course, as is always advisable. Allow me to fantasize further:

Kleppner's dream. One night Professor Kleppner had a vivid dream in which he died, went to heaven, and there encountered Isaac Newton, who proceeded to harangue him to the following effect:

> I see that according to your writings time itself varies as a function of gravity potential Φ, $t = t(\Phi)$, and presumably of motion v^2 as well, $t = t(v^2, \Phi)$. But surely empiricism still and always rules. Suppose, like me, you were curious enough to do your own experiments—and suppose you went into your laboratory and found that strong pressures applied to cesium gas first liquefy it and then solidify it.

Suppose, moreover, that such phase changes, applied to the gas of a cesium clock, were found to alter its running rate. Can you escape the conclusion from this that time itself varies as a function of pressure P? That is, $t = t(v^2, \Phi, P)$? Or, suppose atomic clocks, placed in strong electromagnetic fields F, were found to have their rates affected. Would it then be $t = t(v^2, \Phi, P, F)$? If not, on what basis do you pick and choose between mundane fields and transcendent ones? And, if so, would the possibility of future experimental discoveries leave this parameter called 'time' perpetually an open-ended concept, $t = t(v^2, \Phi, P, F, \cdots)$, so that in principle physical theory could never be completed in respect to this important building block, because the last experiment is never done? Do you see that you are on a slippery slope when once you depart from my original notion about the nature of time, that $t = t(nothing)$?

8.7 The demo problem problem

Would you buy my used flivver? Probably not without a test drive to compare it with your old jalopy. Somewhat the same principle applies to my attempt to interest you in unorthodox descriptive methods. I need to provide a demo problem to show superiority of my proposed new (old) approach. Here, however, I encounter a major obstacle. Convincing demos are not easy to find. Consider three descriptive themes from the past of physics: Instant-action-at-a-distance (IAAAD), retarded distant action (Abraham-Becker, Liénard-Wiechert), and SRT. These seem still in the modern era to preserve an uneasy coexistence, largely supported by myths.

Retarded action rests on the hypothesis that all distant *force actions* are delayed at speed c. For this it seems to lean on SRT and on Maxwell's equations. But ... I have been unable to discern unambiguous support from those theories. Maxwell's equations, to be sure, possess retarded solutions descriptive of radiation, but they also possess advanced solutions (descriptive of what?) and are compatible with Coulomb's observations ... and with Faraday's holistic circuit observations, which provide no empirical support for an inference of retarded action. The entire "theoretical support" for retarded action provided by Maxwell's equations consists in an arbitrary solution-class choice physicists have made. As for SRT, this is a mathematical amalgam of Maxwell's equations and a novel kinematics, strait-jacketed within the unworldly Minkowski space, and otherwise devoid of physics. It really has no special entitlement to any opinion about physical force or its mode of distant action, apart from what it borrows from other schools of thought. The main fact about retarded action is that, as far as I can tell, it is utterly devoid of empirical support. That flivver was bought by the physics community without any pretence of a demo. A vast literature of retarded force action has flowered, supported only by other literature. It will persist for as long as the ignoring of quantum non-locality remains a sound professional career basis.

Concerning SRT, I have already made my quota of critical remarks. Suffice it to say that it is a falsifiable theory, and I have suggested two tests I think it will fail—stellar aberration at higher order than the first, and light-speed measurement in orbit with compensated and uncompensated clocks. SRT has had much successful empirical testing. But, for all its dogma of *spacetime symmetry*, every bit of this successful testing without exception has been asymmetrically on the timelike side (dilation, events of timelike separation)—not a single example on the spacelike side (Lorentz contraction of a structure, events of spacelike separation). In the mind of a sceptic the fact that during a century every empirical attempt to demonstrate SRT's predicted space contortions has failed would arouse suspicion. In that of the true believer it arouses unbounded gratitude for the unending benefaction of repeated trials of faith.

IAAAD has had absolutely nothing going for it for three centuries, apart from its remarkable ability to *save the phenomena* such as gravity, the Coulomb force, and the near field. Ideologically it has never been out of the doghouse. The only reason for tolerating it is that it works. Now, finally, comes quantum mechanics and says that perhaps we should not be parking this rusty old vehicle out in the rain … maybe it belongs in the garage. Maybe we should recognize, reluctantly and belatedly, that IAAAD is quantum mechanics speaking to us in a classical voice. Maybe, by the start of the twenty-first century, we should come kicking and screaming into the twentieth.

I suggest that in a sense you have already bought my flivver. Consider once more the GPS. Let it represent metaphorically the many-body problem, a typical test bed—as it would be, literally, if the satellites interacted. The dynamics of this problem (the *observable* aspect), described in an inertial system, are determined by the nature of the law of interaction. The physics of interaction—the physical evolution of a dynamical system "in time"—is wholly independent of what on-board clocks may be doing ... running fast or slow, showing right time or wrong. Thus the critical choice in trying to use the many-body problem for our test drive has nothing to do with clock rates, light-speed measurements, *etc.*, and everything to do with the assumed nature of the force law.

Basically, this comes down to IAAAD *vs.* retarded action, since SRT is a cipher in this department. So our test problem, to prove the superiority of one path of analysis over the other, would have to adduce empirical evidence favoring one type of interaction law over the other. I am aware of no such evidence. My impression is that no experiment proves retarded action; but by the same token none disproves it. (I set aside evidence of gravity's instant action, which is overwhelming ... without it the solar system would come apart in a few hundred million years. I set it aside not because it is ignorable but because generations of physics scholars have discounted it, while salving their consciences—or, in default of that, their intellects—with "geometry.") In these circumstances no convincing test problem, with a verifiably known answer, presents itself in the many-body area, nor even any practical crucial experiment I can conceive of. Without empirical evidence, one faces a frog-mouse battle of ideologies devoid of physics. And the fact that clock rates do not come into it at all means that my apperception concerning the simplification of physical description (that it is attainable by clock compensations so contrived as to define a CT) has nothing to do with external fact but only with theory and its means of simplification. Simplicity is a characteristic of theories ... so there is no demo problem independent of theory. Although theory cannot prove theory, we like to say that empiricism can disprove theory. But empiricism, alas, sometimes (and usually at the most critical spots) defaults.

And theory, where empiricism defaults, remains ultimately a matter of taste. The upshot is that you will have to kick the tires and decide if the color suits you. But I say you may already have bought my vehicle, apart from a recognition of ownership. For the GPS engineers had little choice in their system design. They had to do it the way they did, because it would be madness to try to describe a many-body swarm of satellites in terms of their individualistic proper times (inexact differentials measured by Einstein clocks). They had to get them all described in common terms employing exact coordinate differentials measured in a given inertial system. If those engineers have already bought the flivver I am peddling—if they have recognized simplicity when they see it—aren't you about ready to sign the purchase papers? The part of your mind that thinks as they do—the part immune to ideology—thinks as I do. Standing in the way is only SRT, with its insistence on the "meaninglessness" (non-invariance) of frame times. I hope to have shown that there is another way of looking at the matter.

A paradigm shift is never easy nor fun. It demands a painful struggle—first within the individual mind, ultimately within the collective mind. Usually it follows upon the impact of new experimental evidence. In this case it must precede the experimental evidence—which never will be looked for without it. Can this reversal of the natural order happen? I know of no precedent. Hence my optimism, such as I can muster, is guarded.

8.8 Collective time in a nutshell

Repetition being the soul of pedagogy, the case for CT can be made in the following condensed way. Consider a swarm of clock-particles in mutual interaction or not. The dynamics of their changing configuration is going to be independent of how it is described. Nature is going to move those objects without our help, each and every one of them and all together … and this is true no matter how we conceive of "time" as flowing or clocks as running. Like the solar system, it all takes care of itself. The task of the wise and lazy spectator, if he wishes to describe this fascinating spectacle, is to find the simplest way of doing the job. I suggest the simplest way may be this:

Find a clock-particle in Galilean inertial motion (uniform and rectilinear with respect to the "fixed stars" or bulk of distant matter in the universe). Treat that as a Master Clock at rest in an inertial frame. Define the proper time of this clock as t_0 and use it as a frame time (collective time) to tabulate chronologically the locations $\vec{r}_i(t_0)$ of each of the particles to be described. (This task might be facilitated by equipping each particle with its own clock compensated to tell collective time t_0.) The resulting record stands in stone, for it uniquely registers historical fact about the system's dynamics—where each particle was located at the moment its on-board CT clock registered t_0. It *describes* what nature is doing ... and since fact is invariant nobody can legitimately claim that a hypothetical non-invariance of frame time invalidates the record. Of course this is an arbitrary way of mapping the terrain, but it has simplicity to recommend it: simplicity owing to use of an inertial system and simplicity owing to use of a common time parameter having an exact differential dt_0, so that the map from one moment to the next has a connected and integrable character.

The way this t_0 is shared among all particles (so that the testimony of each is in harmony with the others) is through a scheme of clock compensations. Patterned on the one put into daily practice by GPS engineers, it requires as information inputs for the ith particle its squared speed v_i^2 and its gravity potential change $\Delta\Phi_i$ relative to the Master Clock. The compensation then alters individual clock rates in accordance with Eq. (8.12), so that the "second" of time means something different for each particle (a different number N'_{0i} of atomic oscillations). This brings their timekeeping properties mutually into step—through shared awareness of CT.

The simplicity then appears in that the simple equation of motion (7.4) [or (8.16)?] governs each particle ... which certainly would not be true if non-integrable proper times were used in our attempt to describe the many-body dynamics. Moreover, the force $\vec{F}_0$ in (7.4) is the simplest form of force—namely, instant-action-at-a-distance, as in Newton's mechanics—the "instant" referring to a given value of CT t_0. (This statement, I freely acknowledge, is an expression of faith, to be tested empirically before acceptance. *Caveat emptor.*) If one wants to know particle proper times, this can be accomplished by individually "decompensating" the clocks (multiplying N'_{0i} by $1/f_i$ to regain N_0), or by letting each particle employ two clocks (compensated and uncompensated). The latter approach would be more practical if it were important to keep continual track of proper-time phases (elapsed proper times or ages), as distinguished from aging rates.

As a reminder: My basic philosophic aim has been to restore simplicity to physical description. I claim to be able to do it with collective time. The contrary notion that such a time is a frame time in the SRT sense, hence non-invariant, hence "meaningless," has no intellectual ancestry except spacetime symmetry, Minkowski space, and the rest of the SRT paradigm, which I hope to have been able in this book to make at least a credible start on discrediting.

8.9 Chapter summary

Space and time are indeed connected, but not at all in the way Einstein thought. We have seen that three-space geometry can be "made" Euclidean by fiddling with clock rate-settings, applying a compensation as an integrating factor to convert non-integrable (proper time) differentials into integrable (frame time) ones. The resulting Galilean inertial transformation equations, (8.17), employ a variant of Newtonian absolute time we have termed collective time. Although CT lacks the physical picture of a reference system or medium in some sense "at rest," it shares with absolute time the highly desirable attribute that its differential dt_0 is *both* invariant and exact, so that particle trajectories are integrable. By the simple expedient of getting all particle-related differentials integrable in the same inertial system, it becomes possible to contemplate a coherent analysis of the many-body problem, complete with Newton's gravity law and his third law in their original, unsullied (instant-action) forms. A vast simplification of physical description results. (This simplification is as yet promissory, since the necessary experiments to establish it have not been done yet. In view of the weight of inferential evidence here adduced, there seems to be adequate motivation to do those experiments.)

Frame time differentials dt_0, as measured by compensated clocks, are invariant under inertial transformations for exactly the same formal reasons as Newton's absolute dt was invariant. On first acquaintance this seems strange, in view of our conditioning by SRT, which insists on non-invariance of frame times, tilting of hyper-planes of constant frame time in Minkowski space, relativity of simultaneity, contraction of space or of material objects in space, absence of "now," *etc.* That whole paradigm is swept away—and good riddance—by the simple expedient of recognizing the objective asymmetry of clock energy (action) state changes and of their effects on physical timekeeping (correctible by compensation). By means of CT, the science of mechanics simplifies formally to its nineteenth-century canonical *forms* (including that of Hamilton-Jacobi, but with "relativistic" modification of mass in the Hamiltonian); the classical Correspondence under-lying quantum mechanics is validated; three-space geometry reverts to Euclidean ... and rigid-body mechanics; centers of gravity, apparent (*i.e.*, as measured by compensated clocks) instant action-at-a-distance and absoluteness of distant simultaneity, action-reaction balance, *etc.*, are reinstated.

Clock compensation amounts (in terms of methodology) to treating the effects on clock rate of known motional energy state changes, as well as gravity potential energy changes, as "apparatus effects" to be eliminated. This makes sense if we are willing to renew our pre-Einsteinian notions of "time" as usefully conceived in the Platonic way Newton thought of it—flowing undisturbed by external influences. In that way it becomes a simplest descriptor. As a dividend, this approach yields recovery of our intuitive notions of "now," past, present, *etc.* Instead, Einstein offered profundity, and physicists bought it ... but profundity comes at the cost of simplicity. How long must this nightmare of professional masochism persist? There is a simple moral here: when someone offers you subtlety or profundity, buy it if you must ... but be prepared to sell it back later—at fire sale prices. The nice thing about science is that it goes on: It's never too late for second thoughts. Religion, by contrast, stands pat on revelation. We have thus arrived at an historic testing point for physics, which (through the political behavior of its priests) will reveal the extent of its religious content.

The *space* geometry useful for dynamical description is evidently controlled as to its basic mathematical character by how we choose to play definitionally with "time." Einstein chose a truly miserable (though entirely "natural") way of treating the latter, which—among its innumerable other faults—shuts the door forever on any simple formulation of the many-body problem, and sets physics back to its early history antedating discovery of the classical canonical group in mechanics. This was by no means entirely his fault ... the *Ur*-fault lay in an inadequate formulation of field theory going back to Maxwell. The error, though corrected formally by Hertz, has been preserved in amber to this day by the inflexible religious doctrine of universal (and lately "manifest") Lorentz covariance

To conclude on another philosophical note: In regard to "the thesis of conventionalism that Euclidean geometry is always a possible mode of formulating dynamical laws,"[8.4] ignorance prevents my holding a brief for any particular school of philosophy (save for operationalism, of which my main criticism is that it has not yet been pursued vigorously enough to have become well-defined as to what it is), and know nothing of conventionalism ... but must say that the present considerations—though not designed with that objective in mind—strongly support the thesis just quoted. It is ironical that to a crew of GPS engineers fell the honor of confirming this airy conventionalist philosophical hypothesis in a remarkably specific, graphic, and startling way. It is equally ironical that all this happened in the real world and transformed the conceptual universe of practical navigation, pathfinding, target location, *etc.*, while placing under SRT a ticking clock-detonator attached to a keg of dynamite, without much notice being taken by any important philosopher or physicist. It shows how remote is the world of theory from the living world.

This book has presented what I believe to be a viable alternative to SRT—something rather formidable to claim in view of that theory's popular and professional image of invulnerability and its supposed now-and-future total monopoly on empirical support. Whether the alternative can deliver—whether it comes closer to physics than does SRT—must be judged by the working out of its implications in the course of time ... as well as by the outcomes of the two experiments I have proposed. On such matters I place my trust in the future. Even as to the logical consistency of what has been unfolded here, only time will judge. I have said repeatedly that the pudding remains to be proven. But I have furnished at least a recipe for it, and have struck a blow for *pluralism* in the foundations of physics. Perhaps this will embolden others to begin thinking for themselves.

Surveying the science scene with as much objectivity as I can muster, I have no doubt that the time of Reformation is coming, though it may not be yet at hand. If the physics monks of today do not want to be driven from their plushy monasteries back into the wilderness of their neglected laboratories—or pensioned off as Henry VIII did their spiritual forebears in England—it might be politically expedient for them to stuff their consensus and acknowledge the possibility of a loyal opposition ... thus to forestall the cruel necessity to open their own minds to a plurality of presuppositions. Such would ensure that the good times continue to roll awhile.

For those readers who skim, it may be a convenience if I finally summarize once again the two experiments I think will defeat SRT and open the way to consideration of alternatives such as that proposed in this book:

(1) SRT's prediction of stellar aberration at second order should be checked, not taken for granted. The VLBI system now has this capability, if its accuracy claims are to be believed. From Eqs. (4.4), (4.5), SRT predicts the aberration angle to be

$$\alpha_{SRT} = \sqrt{1-\ell^2}\left(\frac{v}{c}\right) + \frac{\ell\sqrt{1-\ell^2}}{2}\left(\left(\frac{v}{c}\right)^2\right) + O\left(\left(\frac{v}{c}\right)^3\right),$$

where $\ell = -\sin\theta\cos\phi$ and the angles are as in Fig. 4.1. My prediction is $\alpha = \sqrt{1-\ell^2}\,(v/c) + O\left((v/c)^3\right)$. The latter result would refute SRT, Maxwell's electrodynamics, and the postulate of light speed constancy, and would commend consideration of the Hertzian alternative.

(2) A suitable apparatus for measuring light speed should be placed in orbit in a free-falling inertial system, using a dual-function clock containing counters set to measure proper time and compensated time, as discussed herein. SRT predicts that the proper-time clock will measure speed c, whereas my prediction is that the compensated clock will measure c. The latter result would refute the relativity principle as currently understood, and would commend its reformulation.

Such unexpected results, if observed, might encourage the abandonment of covariant in favor of invariant formulations of the laws of nature, as well as a reappraisal of the meaning of "time" ... and perhaps even of the *modus operandi* that has steered physics into its current impasse.

In the new-old paradigm set forth in this book, have we found a better mask to put on the face of nature? I must leave the question open, letting it serve to test whether *scientific curiosity* is still an active force in latter-day physics. When the experiments have been done, it will be time enough to contemplate the next paradigm shift or **SCIENTIFIC REVOLUTION**.

References for Chapter 8

[8.1]   H. R. Hertz, *The Principles of Mechanics Presented in a New Form* (Dover, New York, 1956), p. 27.

[8.2]   T. Van Flandern, *Meta Research Bulletin* **14**, 57 (2005).

[8.3]   A. Koestler, *The Sleepwalkers* (MacMillan, New York, 1954).

[8.4]   R. S. Cohen, Introduction to H. R. Hertz, *The Principles of Mechanics Presented in a New Form* (Dover, New York, 1956).

Gaudeamus igitur, juvenes dum sumus; post jucundam juventutem, post molestam senectutem: nos habebit humus.::

Ubi sunt ui ante nos in mundo fuere? Vadite ad superos, transite ad inferos: ubi jam fuere.::

Vita nostra brevis est, brevi finietur, venit mors velociter, rapit nos atrociter: nemini parcetur.::

Vivat academia, vivant professores, vivat membrum quodlibet, vivant membra quaelibet: semper sint in flore!::

Vivant omnes virgines faciles, formosae, vivant et mulieres, tenerae, amabiles: bonae, laboriosae!::

Vivat et respublica et qui illam regit, vivat nostra civitas, Maecenatum caritas: quae nos hic protegit!::

Pereat tristitia, pereant osores, pereat diabolus, quivis antiburschius: atque irrisores!::

Gaudeamus Igitur (Old Melody)

Appendix

Dingle's "Proof that Einstein's Special Theory cannot correspond with fact"

After completing the text of this book I happened to pick up and renew acquaintance with the book *A Threefold Cord* by Viscount Samuel and Herbert Dingle (George Allen & Unwin, London, 1961). A fuller account of Dingle's position is given in his *Science at the Crossroads*, but that book is nowadays out of print and essentially inaccessible except at a Russian (!) Internet site. An Appendix II of *A Threefold Cord* reprints Dingle's proof against SRT as "fact." It seems to me this proof has merit, and Dingle deserves belated recognition for his long, lonely display of scientific integrity. Hence I reproduce most of that Appendix below, followed by a short commentary of my own.

Dingle's Proof

> Consider a group of bodies all relatively at rest. Choose one of them as the origin of coordinates, and place a standard clock there. Then every one of the bodies is at rest in a co-ordinate system thus chosen, and the time of any event that occurs among them is obtained by letting a beam of light proceed from the event to the clock at the origin and subtracting from the time which the clock records at its arrival the quantity r/c, where r is the distance of the event from the origin and c is the velocity of light. This is the definition adopted by Einstein, and there is no reason to quarrel with it.
>
> It is often expressed in a slightly different form. We suppose a standard clock, at rest in the co-ordinate system, placed at each point of space and synchronized with the clock at the origin by the dispatch of a light signal from the origin to this clock and its immediate return. Then the clock in question is set so that it records the time of receipt of this signal as the mean of the times, by the clock at the origin, of its dispatch and return. The time of any event is then given directly by the reading of the clock at the place and time at which the event occurs.
>
> It is obvious that this gives the same value for the time of the event as the former definition, and indeed the same

value is given by any clock of this set, if we make the proper allowance for the time of travel of light from the event to that clock. In other words, the time of an event in any co-ordinate system is a property of the *system*, not of any single clock in it. Though this will hardly be questioned at this stage it is necessary to insist on it because many errors have been made in applying the Lorentz transformation (which so far we have not introduced) to actual events, which can be traced to a tacit assumption that the time of an event in a single co-ordinate system varies with the clock, stationary in that system, by which it is determined.

Einstein was perfectly clear on this point. In his original paper on the subject he wrote: 'It is essential to have time defined by means of stationary clocks in the stationary system, and the time now defined being appropriate to the stationary system we call it "the time of the stationary system" '. It is obvious from this that the time, in any single co-ordinate system, of any event has a single value, no matter by what clock of the system it is evaluated. Again, Einstein and Infeld, after giving the definition of time in a co-ordinate system in terms of a multitude of clocks as described above, write: 'When discussing measurements in classical mechanics, we used one clock for all co-ordinate systems. Here we have many clocks in each co-ordinate system. This difference is unimportant. One clock was sufficient, but nobody could object to the use of many, so long as they behave as decent synchronized clocks should.'

It is clear then, that one clock is sufficient, and we shall suppose that that is all that we have—a single clock at the origin that gives the standard time for the system of every event that occurs.

Next, it must be noticed that we say 'the time for the system' and 'every event that occurs', not 'every event that occurs in the system'. An event—a point-instant—is independent of all co-ordinate systems; it is meaningless to speak of it as 'stationary' or 'moving' in any such system. If the event is the instantaneous collision of a large number of bodies, all coming from different directions, it can be regarded as happening to any one of them: they are all moving differently, but the time of the event in any co-ordinate system is exactly the same, no matter on which of the bodies we regard it as occurring. Events are essentially independent of co-ordinate systems, but their places and times may vary with the system chosen.

Suppose now that we have two groups of bodies, all the bodies in each group being relatively stationary and moving with uniform velocity with respect to all the bodies in the other group. We may think of two swarms of stars passing through each other, like the two star streams of Kapteyn, the peculiar motions of the stars in each stream being neglected. Suppose that the stars at the origins pass one another at an instant at which the standard clocks placed on them both read zero, and that at some later time two stars, one in each group, collide. That is an event, and it has a definite time in each co-ordinate system. It is the same event, no matter whether the observer at the origin of stream one regards it as a disturbance of one of his stationary stars or whether the observer at the origin of stream two regards it as a disturbance of one of his stationary stars, but the times of the event which the observers record may of course not be the same.

Let us suppose they are not. Then since the clocks agreed when they were together, they must have run at different rates afterwards, for if we want to compare the rates of clocks the only conceivable way of doing so is to compare the time intervals which they record between the same events. The events belong no more to one co-ordinate system than to the other, but the times that elapse between them are properties of the systems, i.e., of the clocks that record the times of the systems. Hence, if one clock shows, say, twice the time interval shown by the other between the same two events, that clock must go at twice the rate of the other.

Now the clocks are supposed to be of identical construction, so that if one is regarded as running uniformly in its system, the other must also be regarded as running uniformly in its system. Hence, if the *clock-time* intervals in the two co-ordinate systems, between the same two events, are ΔT and $\Delta T'$, respectively, $\Delta T / \Delta T'$ must be a constant quantity; it cannot vary with the particular events chosen for the comparison.

From this it follows inevitably that the co-ordinate differences, Δt and $\Delta t'$, to which the Lorentz transformation applies, cannot, as Einstein's theory requires, represent the clock-time intervals, in the respective co-ordinate systems, between the same events. For, according to the Lorentz transformation,

$$\frac{\Delta t'}{\Delta t} = \gamma - \gamma \frac{v}{c^2} \frac{\Delta x}{\Delta t}, \tag{A}$$

where $\gamma = 1/\sqrt{1-v^2/c^2}$. We therefore get different values for the ratio according to the events we choose, Δx being the space interval and Δt the co-ordinate time-interval, in one of the systems, between those events. We can, by a suitable choice of events, make $\Delta t'$ greater or less than Δt, by any amount we like. But, as we have seen, the relative rates of the *clocks* must be independent of the events chosen for their determination. Hence $\Delta t'/\Delta t$ cannot represent the relative rates of the clocks.

Commentary

Dingle (1890-1978) gives a bit more than this in his Appendix II, but I think it adds little to his argument. I have taken the liberty of replacing his $1/\alpha$ by γ. Here he emphasizes clock *rates*, rightly identifying them as the Achilles heel of the special theory, and frontally attacks the Lorentz transformation, its central citadel. We may observe an interesting evolution of Dingle's thought. For many years he was a believer in Einstein's "strong" form of the relativity principle, which asserts a symmetry of all observables in all inertial systems—hence equality of clock running rates in all such systems. This led him to anticipate (counterfactually) that owing to such "relativistic symmetry" the famous twins must age equally. Yet here he opens his mind to the possibility that this is wrong and that there could in fact be an objective asymmetry of clock running rates ... thus anticipating GPS evidence on clocks, as well as CERN data on muons. The gist of his argument is that, whether such a running-rate asymmetry exists or not, the way clocks are set by the Einstein synchronization convention (which he describes here and which I omitted from my text) ensures that in any given inertial system they will all run uniformly and identically in step, *regardless of where they are situated in space*, exactly as in Newtonian physics. Thus there is no logical room for an x-dependence of timekeeping to arise in any inertial system, nor any physical basis for distinguishing a coordinate origin that would give substance to such a dependence. Yet the Lorentz transformation [(A) above] specifies an explicit position dependence—in fact introduces it at first order, as shown by our Eq. (1.5d).

For comparison, note that our "neo-Galilean transformation," Eq. (7.11a) or (3.23), calls for $\Delta t'/\Delta t = 1/\gamma$, with no spatial dependence. Here t' corresponds to the proper time of a clock at rest in the "moving" system. It plays the same role in the Lorentz transformation, (A), above; yet observe that if the term in Δx were omitted from the latter we would have $\Delta t'/\Delta t = \gamma$, which is upside down. That is, the Lorentz transformation is spared the mistake of predicting proper-time aging greater than inertial-system aging only by the presence of the Δx term. That term saves the theory's bacon. So, it has to be there ... but Dingle's point is that it can't be there: One cannot let clock phases vary with distance without playing hob with the stability of clock rate ratios and running rates of clocks in uniform relative motion: $\Delta t'/\Delta t$ can depend on the constant v^2 (as it does through γ), but not on any variable such as Δx or Δt. The idea of Δt time-variability of $\Delta t'/\Delta t$, as shown by (A), is particularly bizarre. (Think of that: the ratio of two uniform frame-time clock rates, each constant in time, varies with time! Our ancestors who thought Einstein's relativity hard to understand were surely closer to the mark than today's generation that finds it easy.) Whether or not you accept that Dingle was right as a matter of logic, I believe the experiments proposed in the text will confirm that he was right as a matter of physics.

Wherein, then, did the *dementia* of which Dingle was accused reside? I suggest it was in keeping his head when all about him were losing theirs and blaming it on him. Beyond level-headedness, Dingle showed great courage in standing up to the entire physics community. I challenge the physics community to show equal courage by putting its "self-consistent" (paradox-ridden) theory to the test against an equally self-consistent (comparatively paradox-free) alternative.

While I am adding afterthoughts subsequent to writing the text, I should like to thank Bruce Warring for bringing to my attention the book *Popper versus Einstein* (Mohr Siebeck, Tübingen, 1998) by Christoph von Mettenheim. This seems to me a profoundly thought-out critique of Einstein's theories from an unexpected quarter—that is, from a philosophical standpoint. von Mettenheim's approach to *time* bears essential similarities to the one advocated here, but supported by entirely independent arguments. I find it remarkable that by the application of purely philosophical principles a non-physicist (who proves it by making a few minor physics misstatements) can arrive at a deeper and *more practical* conception of "time" than the physicists have succeeded in doing. To illustrate (by exaggeration) von Mettenheim's style of criticism, consider the syllogism:

All foxes have four legs.

Einstein was a sly old fox.

Therefore Einstein had four legs.

The fallacy here evidently lies in analogical thinking, of the sort Einstein himself employed, for instance, when he argued that gravity must act on light in the same geometrical way that acceleration acts on it. The point is that Einstein was not identically a fox, only analogous to a fox. Analogical thinking *proves* nothing: Einstein might or might not have four legs. The only test is empirical: One cannot in logic avoid the obligation to count Einstein's legs.

Philosophers as exemplars of a profession seem more or less supinely to have joined the chorus of Einstein adulation—he being a thinker after their own heart. It is good to find an exception that remorselessly applies analytic criteria appropriate to the philosophic trade. Apparently not all philosophers stand clueless under the aspect of modern physics. In a world of waxing ease and waning comfort, it is a comfort to discover a residue of hope in philosophy ... even though, as the Marxist philosopher Chico so aptly pointed out, *there ain't no Sanity Clause.*

Remark on the psychology of scientific revolutions

In any scientific revolution Nature does not change in the slightest, but the way we perceive our task of describing her undergoes some drastic overturn, of the sort known to psychologists as a *Gestalt* switch. The classic illustration of this is an ambiguous drawing that may be perceived as either a duck or a rabbit. The drawing does not change, but the way the mind chooses to "see" it may vacillate between the two alternatives. The mind can at any moment accommodate one or the other, but not both. If this book has succeeded beyond the avaricious dreams of its author, it will have induced a vacillation in the mind of its reader.

Index

MEMBER OF SCABRINI GROUP

Québec, Canada
2006

Printed in Canada